Penguin Books
Living Through the Blitz

Tom Harrisson left school to serve as zoologist/ ecologist on the first in a series of Oxford University Expeditions – to the Arctic, Borneo and the Pacific (1931-5). In 1936 he was co-founder of the pioneer social research organization Mass-Observation. He directed this until the later part of the war, when he was dropped back into central Borneo. He then stayed in South-east Asia over 20 years (once not speaking English for 10 months) doing zoological, archaeological and occasional political work. He was Director of the Mass-Observation Archive and Professor at the University of Sussex, Senior Research Associate (in anthropology) at Cornell University, Chairman of the Turtle Specialist Group of the International Union for the Conservation of Nature, and Consultant to H.H. the Sultan of Brunei. In January 1976, he was killed with his wife in an accident in Bangkok.

Tom Harrisson

Living Through the Blitz

Penguin Books

Penguin Books Ltd, Harmondsworth,
Middlesex, England
Penguin Books, 625 Madison Avenue,
New York, New York 10022, U.S.A.
Penguin Books Australia Ltd, Ringwood,
Victoria, Australia
Penguin Books Canada Ltd, 2801 John Street,
Markham, Ontario, Canada L3R 1B4
Penguin Books (N.Z.) Ltd, 182–190 Wairau Road,
Auckland 10, New Zealand

First published by Collins 1976
Published in Penguin Books 1978

Made and printed in Great Britain by
C. Nicholls & Company Ltd, Manchester
Set in Monotype Times

To CHRISTINE HARRISSON
who endured three years of domestic blitz (1972–5)
and to the late
RICHARD TITMUSS
never met but profoundly admired for the most
penetrating of British civilian war books.

When I was a young priest, I used to try to unravel what motives a man or woman had, what transplantations and self-delusions. But I soon learned to give all that up, because there was never a straight answer. No one was simple enough for me to understand.

GRAHAM GREENE
The Honorary Consul (*1973*)

It is these faint and separate judgments of probability which unite, as if with an explosion, to 'make sense' and accept the main meaning of a connection of phrases; and the reaction, though rapid, is not as immediate as one is liable to believe. Also, as in a chemical reaction, there will have been reverse or subsidiary reactions, or small damped explosions, or slow widespread reactions, not giving out much heat, going on concurrently, and the final result may be complicated by preliminary stages in the main process, or after-effects from the products of the reaction.

WILLIAM EMPSON
Seven Types of Ambiguity (*1961*)

To the thinking man nothing is more remarkable in this life than the way in which Humanity adjusts itself to conditions which at their outset might well have appeared intolerable. Some great cataclysm occurs, some storm or earthquake, shaking the community to its foundations; and after the first pardonable consternation one finds the sufferers resuming their ordinary pursuits as if nothing had happened. There have been few more striking examples of this adaptability than the behaviour of the members of our golf-club under the impact of Wallace Chesney's Plus Fours.

P. G. WODEHOUSE
The Heart of a Goof (*1926*)

Tom Harrisson with his wife died tragically in an accident in Bangkok as he was returning to Europe to correct these proofs. He would have wished once more to express his thanks to Celia FREMLIN who, as well as her other contributions, took on the task of correcting the proofs and preparing the index, and to Mollie TARRANT who also helped greatly with the proofs.

Contents

Preface

BLITZKRIEG: violent campaign to bring about swift victory.

Pocket Oxford Dictionary

This is an attempt to give an account of the blitz as many British civilians lived through it, often in distress but often, also, in something approaching apathy. It is based on reports written and filed at the time, most of which lay for years untouched, indeed abandoned. They now constitute the Mass-Observation Archive in the University of Sussex, England.

Mass-Observation began, in 1937, as a several-pronged reaction to the disturbed condition of western Europe under the growing threat of fascism. In particular, M-O (hereafter for short) sought to supply accurate observations of everyday life and *real* (not just published) public moods, an anthropology and a mass-documentation for a vast sector of normal life which did not, at that time, seem to be adequately considered by the media, the arts, the social scientists, even by the political leaders.

The immediate response to this idea was massive – both in terms of work offered (nearly all voluntary) and of money. We were thus able to develop a considerable organization. One side of this concentrated on self-documentation, inspired mainly by poet Charles Madge and about the best of the young documentary film men, Humphrey Jennings. The other side specialized in a more observational angle, approaching the study of Britons rather as if they were birds, emphasizing seen behaviour or overheard conversation rather than interviews. This approach derived partly from my own experience as a field ornithologist and member of four scientific expeditions where we had applied methods which appeared (to me) equally applicable nearer 'home'.*

*From September 1939 T.H. became responsible for running M-O as a whole. C.M. engaged in important socio-economic work, and subsequently

The mixture of these two approaches, broadly distinguishable as 'subjective' and 'objective', developed a flow of social information which, alas, was diverted by the Second World War. From late 1939, one had to be in some way useful and 'loyal' to be allowed to continue such studies. M-O was thereby distracted from a purely independent role. To survive, under the security controls and pressures towards uniformity, one had to produce more immediate and 'relevant' results. Our methods, however, remained sufficiently original and adaptable for various governmental and demi-official bodies and individuals to encourage us to deploy a by now nation-wide panel of voluntary 'observers' and smaller groups of well-trained whole-timers. Thus M-O was able to work on a considerable scale through the war. We always retained the right to full post-war use of material, however classified at the time. We did not sign the Official Secrets Act. The result is a large room-full of papers, stacked to the ceiling, which others declare to be unique in scope as well as style.[1]

Although a few post-war authors – Richard Collier, Constantine Fitzgibbon, Angus Calder and Leonard Mosley for instance – have drawn on these papers for their own war books, this is the first 'inside' effort to use the material as a coherent whole. The blitz was selected as a trial subject largely because Philip Ziegler, for Sir William Collins, came down to Brighton, was fascinated by the documents and commissioned us. It has turned out to be about the most difficult, complex and perhaps controversial subject of any among the many larger ones covered by the material. On the other hand, so much has been said and written about the blitz from other points of view, with other conclusions, that the result here may serve to show by comparison what M-O type observation can and cannot do, under situations of exceptional difficulty.

The blitz (a word abbreviated from the German *blitzkrieg*) is generally accepted as beginning in Britain with the first big aerial bombardment of London on 9 September 1940 and thereafter concentrating on that target until a switch to the provinces, via Coventry, on 14 November, and concluding with a record assault back on London 10 May 1941 and a final night over Birmingham

was Professor of Sociology at Birmingham University. H.J. made major war documentary films, and subsequently died in a tragic cliff accident.

on 16 May. The later, lesser 'Baedeker raids' and then the pilotless rockets lie outside our present review. As considered here, a blitz is one where over one hundred German piloted bombers were involved at one conurbation on one night, with a few special cases in addition.[2]

From early September 1940 until far into 1941, M-O accumulated records illustrating the human reactions to being blitzed, both during and after the event, over England, Wales, Scotland and Ulster as the pattern of attack unrolled. The voluntary observers were often, willy-nilly, living in or around blitzed places – and kept on writing. Several hundred kept diaries of candour through this period. In addition, from our 'objective' side, units of trained investigators were sent anonymously to blitz-towns to make overall reports, prepared regardless of any official accounts, departmental feelings or published glosses. (As we shall see, the results inspired strong hostility, up to cabinet level – and little else.) As blitzing was frequently repeated at the same place on successive or closely linked nights, we were present at many actual blitzes, as well as being intimately involved again and again in London. Whenever possible, this writer accompanied the mobile unit; by May 1941 he may have been the most widely blitzed civilian in Britain.

One of the difficulties which became evident, as the present book developed, was that of avoiding the imposition of one's own ideas and accepted attitudes as they had become in the seventies, on to what had been written down some thirty years before. The record from a third of a century ago can seem improbable today because it reads so differently from the contemporary, established concept of what 'really happened' in that war. There has, in particular, been a massive, largely unconscious cover-up of the more disagreeable facts of 1940–41. This aspect, which really deserves another book to itself, is considered again in a final chapter (12). It amounts to a form of intellectual pollution: but pollution by perfume.

It has proved something of an advantage to this writer, co-ordinating and necessarily *selecting from* a mass of old records, that he had an unusually wide experience of living through the blitz. It has been a greater advantage, however, that he has not been subject to the subsequent three decades of brain-washing.

Living, from 1944 to 1970, outside Europe, out of regular contact with British newspapers and radio, he returned to the M-O papers fresh from the jungle and long-house, with a sense of shock. It seemed, after all those Asian years, almost unbelievable that these reports had been written or (more often) read by this same person in 1940 and 1941. This impression of total freshness helped put in proportion some factors from the records (though perhaps it unbalanced others).

In order to recapture the atmosphere of those remote nights and to unstick those myriads of unremembered words, a sort of psycho-refreshment was at times found helpful. This effect was achieved occasionally by sitting, writing, in London tube stations; or more often in the long out-patient corridor of the great Hospital of St Pierre, Brussels, which must be one of the busiest arteries of human anxiety, distress and relief.

M-O's relationship with its voluntary observers and diarists was based on an assurance of their anonymity. This enabled them to write with complete honesty, even in terms that suggested defeatism, panic and other 'unpatriotic' reactions. That anonymity is, of course, still honoured here. But the full-time paid staff worked under no such seal; several of these, who made major contributions, are described in the chapter notes where relevant, along with some specialist historians and others who advised generously on special points. In the past some critics have tried to discount the whole of M-O as leftist or dilettante. The records of the men and women, active in the work both then and since, refute this fallacy. If any such charge is repeated in the fourth quarter of the twentieth century, it may only be because some Britons, especially responsible ones, cannot face the full facts about their 'finest hours'.[3]

One of the war's official historians, P. I. Inman, in a preface to *Labour in the Munitions Industry* (1957) noted, rather sadly, the limitations of having to write only from ministry files and other official sources. Visits to factories helped, this author found, 'for the vitality and significance they gave'; but 'such visits make the historian aware, if but dimly, of the infinite variety of individual knowledge, thought, anxiety and toil that lay behind war production – of the reality which his ordered narrative can never capture'. Yes, this is a real headache: how to recognize and to report justly that 'infinite variety' while keeping order in the narrative – that order which readers have come to expect from history. For better or for worse, we can only insist that *every* generalization about forty million Britons is almost endlessly subject to qualification. At no time in the Second World War generally and in the blitz particularly were British civilians united on anything, though they might be ready to appear so in public on certain issues.

This account tries to preserve the marvel of that human variety, although to have played it down would have made it much easier to draw conclusions with impressive sweep. Samuel Johnson saw it another way, back in 1758, when he could not decide 'whether more is to be dreaded from streets filled with soldiers accustomed to plunder or from garrets filled with scribblers accustomed to lie'. To find and then to spell out the 'truth' about the blitz and its victims is difficult indeed, despite the wish to do no less.

Warm thanks are due to Mrs Mary Adams, Professor Asa Briggs, James Fulton, Dr Paul Addison, Richard Fitter, Jo Sinclair, Alastair Maclean and Ian McClaine for special help in England; to Stephen Spender, Naomi Mitchison and Kathleen Raine for allowing me to quote their poems; to the authorities of Belgium's École de Guerre library, and their Ministry of National Education's Centre for Research and Historic Studies of the Second World War (both in Brussels) for intensive reading facilities, mainly useful in identifying themes derived from the British material and comparing these with the experiences of other nations.

In trying to spell out this story, great and generous help has been received from Celia Fremlin (Goller), who prepared a first draft of chapter 5 and helped with other, wider ideas. She was herself one of our most energetic wartime observers. But the final version – in so far as it is final – with numerous faults partly implicit in the record itself and partly in the author himself, is the responsibility, subjectively and objectively, of

TOM HARRISSON
Brighton and Brussels,
1 November 1975

1 The Expected Holocaust

Tsar Semon, expecting to fight as before, marched to war against the Indian Tsar; however – what worked once does not always work twice. The Indian Tsar did not even allow Semon's army to open fire, but sent his women through the air to hurl explosive bombs on Semon's army. The women began raining bombs on Semon's army from above like borax on cockroaches; Semon's whole army scattered and left Tsar Semon alone. The Indian Tsar took over Semon's realm, while Semon the Soldier ran off wherever his legs would carry him.

LEO TOLSTOY (1828–1910)
The Tale of Ivan the Fool

Air-power was born with a kind of huge attachment, a Siamese twin, so that the facts of the matter were intimately blended with an almost separate, quite identifiable other form called Fear-Fantasy. This second fellow grew up alongside brother Fact, receiving some special stimuli in the First World War (1914–18) and continuing over the twenties and thirties of this twentieth century as, among other things, the symbol of an ultimate human disaster.

The idea that attack from overhead would become the final, totally devastating stage in coming wars grew, indeed, to cloud almost all thinking on this subject. It became near-obsession – comparable, say to the one-time belief of strict Christian sects in a burning hell for the unredeemed. The pattern of British politics and forward planning was gradually overshadowed by visions of shattering bombardment on the civil population. Along with this, latent where not explicit, went the broad assumption that much of this population would either be killed, shell-shocked or reduced to panic. They would not be able to stand up to the experience. Thus the properly motivated, disciplined armed forces of traditional war would be threatened by collapse on the home front, fatally stabbing them in the back.

The growth, then fixation, of this grim twin was not simply fed on the private alarms of responsible politicians, nor the latent death-wish of some wider group, seeking the apocalypse. This fearful concept of coming 'reality' was supported by scientific argument, statistical estimates of the highest order. Predictions came from extreme left as well as furthest right. The foremost international theorist of the theme was a fascist general in Italy. The foremost alarmist in Britain was the country's leading communist scientist.

The conflicts between fantasy and reality are one of the main subjects of this book. To understand what happened when Britain was heavily bombed, we must first recognize that much of the suffering that resulted derived directly from this earlier thinking which dictated what preparations were made, precautions taken, things done and left undone.

The later part of the First World War gave Britain its first experience of bombing. The experience made a great impression on – among others – the then budding young politician, already a minister, who was to lead his country through the second, much greater experience of the Second World War.

During the first war, Germany sent some 200 airships and 430 aeroplanes over Britain, producing 4400 casualties, nearly all civilian. The most important incident occurred on 13 June 1917, when seventeen twin-engined Gothas dropped about two tons of bombs on London, mostly around Liverpool Street Station. 162 people were killed, without German loss. A similar raid on 7 July killed 75.

In those days there was no accepted system for reporting mass reactions objectively. Public opinion was commonly generalized from above; usually manipulated or misrepresented by mistake. Plenty of statements declare that 'the citizens of London were outraged by the bombing'. A. J. P. Taylor uses the adjective 'hysterical'. There certainly were strong responses in the crowded East End of London, largely (it appears) because of the absence of any protective measures from either the military or civilian authorities. And there was then no Labour Party, let alone Coalition government, to let the 'poor' feel represented at the top.

Impossible as it is now to be sure of the public mood then, the voices of leadership are on record, loudly. 'Parliament was in an uproar.' Government was momentarily shaking, some parliamentarians at once foreseeing the implications for a greater air-war. Wing-Commander Allen, in a critical study of air-tactics generally, perhaps overstates his case when he concludes that the London incidents were 'to result in a massive reorganization of Britain's armed forces which would affect her strategic doctrines for the next fifty years or more'. But at the least here was the start of a trend which led the island's leaders to plan elaborate strategies, civil and military, aimed at avoiding anything of the same kind in the future.

At the same time, the voice of government spoke bravely, reassuring the people then as in the following war. The Minister of Munitions, age 36, made the first of his many dicta on the subject: 'It is improbable that any terrorization of the civil population which could be achieved by air attack would compel . . . surrender . . . we have seen the combative spirit of the people roused, and not quelled, by the German air raids.'

Churchill went on, in 1917, to apply the same argument, logically enough, to German civilians. 'Nothing we have learned,' he declared, 'justifies us in assuming that they could be cowed into submission by such methods.' Some have since puzzled over what 'changed his mind' so startlingly by the forties, when he not only allowed but encouraged a vast deployment of the nation's war effort precisely into seeking that submission, at the cost of 59,223 air-crew killed in RAF Bomber Command, compared with 51,509 civilians killed by the Luftwaffe bombing of Britain.[1]

To some extent Churchill did change his mind; but the change was nearer a shift, small, easy. For it rested on a deeper belief in the weakness of the common man. Many Conservative and Liberal leaders never trusted the masses and in a way deeply, privately, despised them. They could easily visualize reds under the bed, ready to crawl out over the soft counterpane. The 1917 Russian revolution from behind the lines had set a terrifying precedent.

But Churchill, a pugnacious pragmatist, was no deep theoretician. The influential theorist of mass demoralization was the Italian General, Guilo Douhet (1869–1930), who developed his ideas in his *Il dominio dell' aria* (1921). Gazing into the Mediterranean haze he foresaw wars fought by droves of bombers which would render armies and navies redundant, aerial defences miserably ineffective. He based his arguments largely on the London attacks of 1917 and the subsequent successes of Italian planes, flying over 600 miles.

Douhetism implied (where it did not spell out) almost random assault on the home front, a clear overlap of armed forces, paralysing the social and economic life of the enemy. He urged: 'hammer the *nation* itself to make it give in'. 'It will,' he added, 'be an inhuman, an atrocious performance, but these are the FACTS.'[2]

Mussolini publicly adopted this view in a speech to the Italian Chamber of Deputies, June 1927, when he bellowed that his nation *must* have an air force 'numerous and powerful enough to stifle by the roar of its engines all other noises in our peninsula'. With polluting poetry Il Duce envisioned 'That the span of its wings should intercept the sunlight'. Field-Marshal Jan Christian Smuts, the Boer leader with much influence on Churchill, experienced the 1917 London raids and was appointed a special adviser to the War Cabinet on home defence measures. He and Trenchard became prophets of the bomber's approaching omnipotence. Trenchard was to become its most active protagonist.

Adolf Hitler had a less educated view of war. The Führer did not encourage a heavy emphasis on the bomber. The accidental death of Germany's leading Douhetist, General Walter Wever, helped lead to the pre-war cancellation of plans for heavy bombers in favour of speedy planes, fighters, short-range dive-bombers. Only reluctantly did more desperate pressures send the Luftwaffe 'indiscriminately' bombing. Then, late in 1940, their continuing bomber weakness enormously alleviated the burden to be borne by British civilians, whereas Trenchard's enduring vitality had, eventually, the obverse effect upon the unhappy inhabitants of the Third Reich.

Mussolini's early enthusiasm was fortified by further experiment upon the Abyssinians, Trenchard's by RAF attacks upon similarly placed Arabs in the Middle East in operations which involved the young Portal on his way to fill Trenchard's original post of power, Chief of Air Staff. By May 1928 Trenchard could confidently circulate a memorandum proclaiming the future unique air-role: to 'pass-over' the enemy defences and assault forces. As he disarmingly concluded: 'It is simply that a new method is now available for attaining the old object, the defeat of the enemy nation.' The bomber would do this on its own. 'Once a raid has been experienced [his long paper says] false alarms are incessant and a state of panic remains in which work comes to a standstill.' The later part of this prediction was to continue an article of bombers' faith until long after the blitz on Britain, regardless of the results in Coventry, Birmingham, Hull.

Although for a while the chiefs of the army and navy questioned these claims for air omnipotence, the Trenchard presentation continued to gather momentum; with important side effects not only on service planning but on a wide arc of responsible thinking, defensive as well as offensive. For, clearly, the argument could be applied both ways. Just as in 1917 Churchill had declared German morale to be as unflinchable as British, now British became as vulnerable – even unto panic – as German. *No one* was visualized as surely able to stand up to terror from the skies.[3]

So it came to pass that on 19 May 1941, in one of the war's odder operations, Churchill himself circulated a long memorandum (he detested these normally) from the now retired Lord Trenchard, 68. The last of a long, almost continuous series of night bombing on British cities had occurred only three days before, 160 tons of HE on Birmingham. The blitz proper of the present book was in fact over.

After a warm tribute to 'the *ingrained* morale of the British nation' as evinced by this long experience of bombardment, Trenchard goes on to reverse his 1917 argument and subhead his second paragraph: *History has proved that we have always been able to stand our casualties better than other Nations.* This reversal was essential to allow the rest of the argument: for the production of many more heavy bombers to shatter – presumably – uningrained morale of 'the civilian front behind the German lines'. Indeed 'all the evidence' now demonstrated that 'the German nation is especially susceptible to air bombing'. This time, the Chief of the Imperial Staff, Chief of Naval Staff and

Chief of Air Staff (Sir Charles, to-be-Lord, Portal) were unanimously in agreement, offering only minor modifications. Churchill readily concurred.[4]

Both parties to the Second World War thus devoted huge energies to the destruction of what they thought to be the other's civil structure and human will. In this massive application of theory – knocking the stuffing out of the masses for victory – they thus severely dislocated most of the main concentrations of civilians in north-western Europe. As late as 14 February 1945, one of the most devastating attacks smashed Dresden in south-east Germany, killing more than in all Britain's blitzed cities put together. The reason, as a young historian suggests, 'seems to have been mainly that Sir Arthur Harris, faithful to Douhet's theory, was determined to let loose one more major air bombing attack'. 'Bomber' Harris, heading Bomber Command under Portal, continued to believe he was winning the war that way.[5]

A continuing, deeply ingrained contempt for the civilian masses had major effects, too, on preparations for the receipt of the bombs at the British end. The proletariat were bound to crack, run, panic, even go mad, lacking the courage and self-discipline of their masters or those regimented in the forces. The alternative view – that a Belfast plumber might manage as well as an Irish Guards officer – was rarely seriously advanced.

The predictions for the civilian were therefore uniformly black. By 7 February 1934, Churchill had warned the House of Commons of 'the cursed hellish invention of war from the air', changing everything. Later that year, on 30 July, he described London particularly as 'the greatest target in the world, a kind of tremendous, fat, valuable cow, fed up to attract beasts of prey'. A week later his most trusted adviser, Professor Lindemann (later Lord Cherwell) wrote to *The Times* warning that bombs 'might jeopardize the whole future of Western civilization'. On 28 November 1934 Churchill again in the Commons elaborated this theme. He predicted that, when London was bombed, three or four million civilians would flee the city. The whole 'vast mass of human beings' would lie, cow-like, helpless and hopeless. They would probably render the army helpless too, as it would have to restore order out of 'the immense, unprecedented difficulties' ensuing.[6]

The official Conservative leader, Stanley Baldwin, had already warned in famous words: 'the bomber will always get through'. Pushed by Churchill and others, in October 1935 he increased warning pressure, and later foresaw 'tens of thousands of mangled people – men, women and children – before a single soldier or sailor' suffered a scratch.

Not to be outdone, Labour's leader, Major (from the First World War) Clement Attlee, on 28 October 1935 spoke for the Opposition: 'We believe that another world war will mean the end of civilization.' Fear spread through all strata of official life. It worked in two main directions. First, linked with increasing alarm at Germany's growing power and the wish to placate Nazism, it caused government to seek for international air disarmament. Second, it conditioned the nature of the preparations made to receive the 'enemy', which were widely based on the 'chaos' concept so ably expounded by Churchill in particular.

In making these preparations, grounded in such fears, the government had naturally to try to work out what it was really going to be up against. The vague holocaustic visions formed a background to such planning. More to the point however, were theoretically more precise predictions based on calculation and mostly secret statistics.

The special cabinet committee set up under Smuts as a result of the first war bombing was the first of these official predictive sources. In 1917–18 the Germans had staged 103 raids (51 by airship), mostly on London, their 300 tons of bombs killing 1413. Deep concern at the deductions drawn from this led, through devious official channels, to a special Air Raids Precaution (A R P) sub-committee in the Home Office, which first met in May 1924 and continued into the thirties under that most administratively able of chairmen, Sir John Anderson, later Lord Waverley, Churchill's trusted wartime minister and patron saint of the 'Anderson shelter'.

The emphasis, from the start, lay on loss of life, coupled with the assumption that London would be the prime target from the first hour of hostilities. By 1924 the Air Staff were offering Anderson evidence that London would receive 450 tons of bombs in the first 72 hours, killing 3800 and seriously wounding twice as many. In the first month the capital would have over 25,000 dead. This sounded pretty grim in the early twenties. But much worse was to come. By 1937, the Air Staff had greatly stepped up expectations, while retaining an earlier casualty figure of nearly 50 per ton of bombs, a figure without statistical validity which nevertheless became accepted as gospel by senior civil servants.

Thus, when the Committee for Imperial Defence in 1937 worked on compensation rates for casualties, it seemed normal to start with 1,800,000 in the next war's initial two months, a third dead, at a cost of £120,000,000 (1937) money. For the following two years, into real war, the Ministry of Health, following the same code, got up to 2,800,000 hospital beds needed for casualties.[7]

Capturing this numerical mood, the Home Office worked out that 20,000,000 square feet of seasoned timber would be needed each month to provide coffins. As this figure was unacceptable, the answer would have to be mass graves and burning with lime. Similarly the cabinet in 1937 decided it would be impossible to insure against enormous air damage (Lloyds had already refused to handle policies). They set up another special committee to study the problem. The gentlemen – women were practically non-existent in between-war planning despite the domestic issues involved – adopted a material damage multiplier of £35,000 a bomb. They concluded that 5% of all property, valued at £550,000,000, would be destroyed in Britain during the first three weeks of hostilities. The cabinet accepted this estimate (October 1938) and it must, in itself, have weighed like an albatross of anxiety round Chamberlain's neck.

By then, in 1938, the estimate for bombs upon London in the first day and night had been upped to 3500 tons, with 700 tons daily thereafter, delivered with high accuracy, mostly by daylight. The theory of a knock-out blow on the capital influenced all these plans and, as Dr Richard Titmuss later judged (1950) 'explained much of the birth and development of the wartime emergency services'. These services – as we shall see – put their emphases overwhelmingly on death and destruction, or crippling hurt, with little consideration for other, less obvious and in the event usually far greater side effects on humanity: confusion, anxiety, dislocation and distress.[8]

Well, in those pre-psephological days the noise of the general public, as interpreted by the media, could sound very different from the true, private, voice of the people, which might be saying the opposite – or nothing at all. There were plenty of public published voices anyway: journalists, scientists, participants in the Spanish Civil War, armchair experts. Lord Halsbury, for instance, forecast a single gas-bomb killing everybody between the Serpentine and the River Thames. 'Experts' consulted by best-selling reporter Negley Farson predicted 30–35,000 casualties a day. A group called the Air Raid Defence League saw 200,000 casualties in ten days.

Liberal leftist John Langdon-Davies, war correspondent in the 1938 Barcelona raids, heavily stressed 'the technique of silent approach' by glide-bombing. This was likely to be 'the normal experience of London'. He doubted if anti-aircraft guns could be used during raids, since the noise would shatter normal eardrums. Powerfully supported by the Left Book Club, leading physiologist and statistician Professor K. B. S. Haldane flatly proclaimed in 1938: 'I would far rather be in central London during a big air-raid than in a traffic jam on the Barnet Bypass.'[9] The eminent scientist expected organized machine-gunning of refugees as they fled from the cities to block the arterial roads. The requirement here, he argued with characteristic force, was to reserve these roads for omnibuses only. These should be spaced '50 or so yards apart, protected by British fighting planes'. He, too, saw sound as a decisive factor. His description of bomb blast is horrifying enough; but then follows 'a sound like that of the last trumpet which literally flattens out everything in front'. Those fortunate enough to survive this sound-wave in anything like one piece can thereafter live in perfect peace. For 'it is the last sound' they'll ever hear, their eardrums burst inward, 'deafened for life'.

Professor Haldane, a dedicated and consistent (rich, upper-class) communist, prophesied panic as thorough as that of any Douhetist. Being on the proletarian side of the road, he spelled it out without evasions. 'Brave men ... find themselves running.' Priority must be given to crowd control: 'anti-panic measures are of the greatest importance'. As regards lethal effects, however, this top scientist settled for a moderate 20,000 dead on an average 500-plane raid, although with a first raid on London obliterating up to 100,000 Cockneys as an opener.

Verbally bombarded left, right and centre, the British public listened anxiously – or turned a deaf ear. Dr Titmuss refers (in the opening chapter of his war history) to 'the general tone of public thought, strongly affected by self-styled strategists in military matters, by the prevailing political climate and by the current of world events'. How far was the public really taking in all this talk?

In the thirties, objective study of public opinion was in its infancy, analysis of private opinion and everyday behaviour embryonic. One of the first to probe into these hitherto obscure sectors of common life was Mass-Observation, starting in 1937. The evidence, albeit incomplete, from this and other early sources of the kind shows, sometimes strikingly, that the people maybe heard much, but by and large *heeded* only a little. Few took the grim published messages either wholly or lastingly to heart. Partly this was due to a chronic cynicism about utterances from above, well established then (and perhaps ever since democracy began); partly it derived from a mix of fatalism ('if your number's on it'), apathy ('what can *I* do?') and 'wishful thinking' by which so many live at any time. At the extreme, however, in 1938, under the shadow of Barcelona, a surprisingly strong intellectual faction seriously considered the possibility of killing off their own families rather than let them suffer aerial war (as Mass-Observation's 1939 Penguin Special *Britain* showed). In fact, no one ever did anything as drastic as that. At the other extreme, many decided to do nothing in advance; they likewise had, in due course, to act differently from their declared intent. So many of peace's anxieties were eased with war, so many more increased; it was the same in reverse with the apathies.

By the start of September 1939 many Britons were expecting a war of some sort, though not necessarily a holocaust. Nevertheless a large bloc remained convinced there would be nothing of the kind, so that they were in a sense taken by surprise when war followed within days.

During the eighteen months before this war began Mass-Observation tracked the growth or, more exactly, flow of opinion on this crucial point of expectation. For it *was* crucial, then. If people did not believe in the predictions as, at least, probabilities, they were unlikely to react positively to official instructions, to prepare air-raid shelters or black-out their homes properly, let alone to train as Air Raid Wardens or join other voluntary services.

British leadership had not led its electorate clearly up to this stage. Back in August 1938 about one in three adults said they expected war, as signalled by Hitler's occupation of Austria, then Czecho-Slovakia. A larger number, however, expressed positive views *against* there being a war. Many more, especially females, were vague, uncertain, or 'uninterested' in the whole subject. Moreover, well over half even of the expectant third did not think trouble would come as soon as it did; they thought of 'after 1939', often long after, something remote, cloudy, requiring no early response in the way of preparation.

This mood did not strikingly change in the year following. And it affected mayors as well as miners. As late as the end of August 1939 a good majority continued to reject the likelihood of war; very many thought Hitler was bluffing. Wishful thinking was rife ... the attitude was expressed by a workman in the Threadneedle Street subway:

Hitler's got Danzig, but is he going to hold it?[10]

For most, the truth only dawned in those first days of September 1939. In the little lull between the German invasion of Poland and Chamberlain's declaration of war, Mass-Observation's diaries are filled with minor gloom, lit with dwindling flashes of hope; 'relieved that something *definite* has happened', too. In the streets, random conversations and questions give as much as half the voices saying they would rather get it over with now. At the limit of tolerance, a Lancashire lass admits: 'My nerves have completely gone; we've been waiting a whole year, not knowing if there'll be a war or not. I want a knock at Hitler.' Or a Cockney: 'We might as well have it and finish with it.' Millions of people entered the Second World War unprepared on the physical level for the predicted worst. That is not necessarily to say, though, that they were equally unprepared psychologically. Indeed, the two levels of self-preparation (insofar as they can be separated) were to some extent contradictory. Stated disbelief in an imminent war with all those implications of the 'destruction of civilization' did not necessarily mean that those persons had not deeply thought into, worried out, been through their personal fears. On the contrary, many seeming optimists had worked the danger over more thoroughly than others who directly, phlegmatically, accepted a fatal certainty. It was perfectly possible to *decide* to do nothing; to suspend judgement and wait until the authorities stopped talking and instead directed *everyone* to do what was necessary. It was up to 'them'. What had 'they' planned?

2 Air Raid Precautions (and People)

Your agony, England!
Arise, rose of the blood,
Blaze from thorns of the heart!
Martyred blood of garlanded past days
Burn from the bombed and bitter island! . . .
Covered with ashes is your statued past,
Once sunlit, where each soul his heritage
from-birth-to-death-span tilled, plotting deliberate choice
Of his willed good or evil flowers.
Destiny more glittering and vast
Than simple single sight facing through life to growing death and knowledge,
Blinds faith's towers.

STEPHEN SPENDER 1942

The directions to be given to millions of civilians so that they might survive these expected tribulations had been the subject of grave concern, prolonged thought, at the highest levels of central government. It was the clear duty of leadership to plan for the worst. The main assumptions were necessarily so based.

The blanket term for this preparatory planning was Civil Defence. The more particular area of civilian concern began to be spelled out by Sir John Anderson's original Air Raid Precautions Committee of 1924. The dedicated, clever civil servants concerned were faced with two difficulties. They were, first and foremost, bombarded by dire projections coming down in a darkening stream from the Air Staff, so that the best they could do was plan 'with a view to mitigating, *so far as possible*, the evils attendant upon aerial bombardment'. 'Ministers,' as Churchill wrote later, 'had to imagine the most frightful scenes of ruin and slaughter.'[1]

The second great difficulty was one of human understanding:

establishing how the masses out there beyond Wigan Pier could be best prepared for these coming terrors. Men like Sir John Anderson, concerned men of the highest intellectual quality, were far more remote from 'the people' than is readily realized today. That made it all the easier to fall back on the Douhetic answer. As Titmuss found, in post-mortem, the planners 'accepted almost as a matter of course that widespread neurosis and panic would ensue'. Thus on the one hand they were led into overestimating the material impact of German attacks, while on the other their own upbringing and outlook caused them to underestimate normal human capacity ('courage'). Under these difficult circumstances they did wonders. Another official historian, Terence O'Brien, summarizes their achievement as 'in many respects masterly'. That adverb is appropriate in two ways, however; master-and-man attitudes underlay the whole operations.

The master plans that resulted could only be tested by events. Before seeing how these events worked out in human terms, it is necessary to look a little more closely at what was prepared as Air Raid Precautions inside Civil Defence, in so far as this was intended to affect human behaviour under bombardment.

It must be emphasized that the standard prediction always saw the capital as first and prime target; to some minds, the only one that mattered much. ('Churchill's sentence quoted above ends with the words "in London".') This was the more natural since the key ministers expected to be implicated – the Home Office with a new Home Security twin, Health, Food, Transport, Works, Labour, Pensions (and Assistance Board) – were Whitehall-based, with Westminster for political pressures along the street. It was also easier to 'keep in touch' with Londoners, there were decision-makers on the doorstep. Metropolitan experience could be shared, pooled, reshaped at all levels quickly, when Cabinet Minister and Poplar housewife would suffer the same attack. Moreoever, the largest London administrative unit, then the London County Council, was headed by Herbert Morrison, to become second Minister of Home Security in war.

In the 'provinces' this set-up was diffused. The processes for decision and revision were far away – in space and in organization. Each separate city, borough and smaller authority was nearly autonomous. Foreseeing the weakness of this traditionally strong local authoritarianism when tested by violent air (or land) attack, a system of twelve Regional Commissions had been established under Home Security, to come into effect with the outbreak of war. Set up amidst heavy secrecy, the regional system was first publicly revealed in February 1939. It had been tested on paper in the Munich crisis – and at once found to raise the hackles of many local bodies, who felt their ancient, locally elective roots threatened by a new form of 'dictatorship'. The regional scheme came into war diluted. The Regional Commissioners, elderly gentlemen of distinction, held sweeping emergency powers in theory; but these too stemmed from the holocaust concept. They did not come into force until and unless the local authority had admittedly 'collapsed'. Although plenty did indeed collapse for short periods, not one admitted it. The R C's had no authority to intervene directly except in extremity. This extremity was defined as complete disruption of communications *with Whitehall.* That, too, never occurred.

In consequence, though the R C's contributed a lot in advisory, co-ordinating or support roles, when the blitz came they proved to be paper leaders. Often, however, there were no others, for significant periods.[2]

The *immediate* response to the bomber had to be to limit what damage it could do. A prime aim was to prevent deliberate aiming on key targets; which inevitably meant bombs scattering over surrounding homes and countryside. This was the primary role of the military (anti-aircraft and balloon barrages), the night fighters and radar. These, as Baldwin said, never could stop the bomber getting through. But the feeling that there was a major effort at defence – the sheer noise of it – enabled most civilians to avoid a sense of helplessness, of total vulnerability, which is one quick fertilizer of panic and which was probably behind the most disturbing reactions to air-attack noted elsewhere: in East London, 1917, then Arabia, Abyssinia, Spain, and later northern France in 1940.[3]

Once the defence of life, property and industry had been taken care of, the rest was up to the civil side, A R P. For those who could be spared from the cities, evacuation was the official first answer.

Evacuation was designed 'simply and solely as a military expedient, a counter move to the enemy's objective of attacking and demoralizing the civilian population' (Titmuss). It was an *organized* mass exodus before the event. Another special subcommittee from the Committee of Imperial Defence started thinking along these lines in 1931. They considered evacuation primarily 'as a problem of preventing panic flight' by removing the supposedly most nervous from danger, with special concern, of course, for 'an orderly exodus from London'. By 1937, the Home Office was reacting to the Air Staff's increased forebodings, deciding that their new estimate of death 'strengthens the case for non-essential persons ... so as to lessen the danger of panic and stampeding'. By 1939 the scheme to move primary school children, younger children with their mothers and expectant mothers was finalized and extended beyond London to include Clydeside, Tyneside, Merseyside, the industrial midlands and in the south Portsmouth and Southampton but not Plymouth, Bristol and Cardiff.[4]

So, in the first days of September 1939, a million and a half persons were evacuated officially, though on a voluntary basis: two-thirds of those qualified went on Merseyside, a third in Portsmouth and Southampton, only 20% in Coventry, irregularities partly due to very uneven local information and leadership. A larger number of all sizes and shapes evacuated themselves, mostly into the west country; an estimated 2,000,000 townees, though – astonishingly, perhaps – there are *no* official statistics. About 'private evacuees, the historian knows nothing'.

By the act of evacuation, several million Britons removed themselves from the immediate fear of bombing. When the bombs were slow to fall, many returned; over half in Manchester and Sheffield, a third in Greater London. The bombers came, after one yeaı of war, and some of these and others too went back into the country, fewer than the first time, though perhaps 1,250,000 humans spread over months. Whatever the exact position numerically, negatively the advance exodus may be presumed to have removed many of the more nervous, adults included. Did the 'cowardly' go? We shall never now know. On the evidence of those who stayed, the reasons for coming and going were multiple, complex, highly varied and variable, not justifying simple classifications of fear or composure, forethought or apathy. This complexity is bound to emerge as a main theme in what follows.

It does seem reasonable to suggest, though definitely not to assert, that the evacuation did ease urban burdens, though it may be impossible at times to detect any difference in end-result between a theoretically evacuated town like Portsmouth and one not allowed for in any scheme like Plymouth. In both, as in other places, unplanned, uncontrolled evacuation could still suddenly develop, especially in the provincial form which came to be known as 'trekking'.[5]

The birth of A R P proper – including precautions for the unevacuated urban mass – can be fully identified in 1935, when Baldwin's Cabinet approved £100,000 for expenditure to promote serious planning. Wing-Commander E. J. Hodsoll, who had been secretary of earlier committees since 1929, was appointed head of a separate A R P department in the Home Office, continuing the approaches begun by Anderson in 1928, with emphasis on the delegation of authority through ministries down to local authorities. Hodsoll was raised to Inspector-General of A R P by early 1939, with a working budget up to £830,000, plus over £6,000,000 for gas masks and other equipment. But a much wider range of responsibilities was left with other ministries and with the local authorities.

250 local authorities, in all shapes and sizes throughout the United Kingdom, were made responsible for nearly all precautions affecting people. Hodsoll and his colleagues had to think and plan. Central government had to help find, deliver and inform (e.g. Health Circular 1779 of 28 February 1939 on recommended arrangements for war burials, followed in April by a first distribution of 1,000,000 funerary forms). But implementation was the task of the individual A R P controllers. Nearly half of those (120) were Town Clerks or equivalent officers, the paid heads of the local civil service. More than a quarter (70) were Chief Constables heading the autonomous local force. The rest were locally elected politicians, Mayors, Aldermen, Councillors. Once war began, these A R P Controllers were to be far and away the most important individuals in every raid outside London. They ranked over

other main service chiefs, including the chief warden, the chief medical officer, the borough surveyor. Although police forces and fire brigades remained under their respective heads, they had to be co-ordinated by the controller. The character, capacity and experience of a controller was likely to be crucial to the welfare, if not the survival, of a whole community. But none of them was selected, appointed or even intensively trained especially for the job.

Among those supposedly controlled by controllers were, presently, over one and a half million civilians required as the planned 'fourth arm', the fully civil ARP side of Civil Defence. By the end of 1938, stimulated by the Munich Crisis, 1,400,000 adults had volunteered – there were still a million unemployed at this stage. This fourth army, to cope at the first stage with all sorts of bombs, gas, casualties, shelter-control, etc., had specialized sections including Rescue, First-Aid, Ambulance, Communications, and Decontamination (poison gas) grouped on a scattered system of Wardens' Posts, which developed varied infrastructures of their own, best reflected in the hundreds of magazines they began to produce long before the bombs fell, with titles like *Arpoons*, *ARP Sauce*, *AR Pie*, *Blackouts & Warbles.* These featured Hitler jokes, locally topical cartoons, serious raid advice and poetry from Southgate:

> It's hard work lifting a stretcher
> You betcher
> You'll be dead if you injure your aorta
> Saorta!

And from Middlesbrough in the north:

> In a raid if you must lose your head,
> Just remember the things that you've read.
> You'll know what to do;
> There'll only be two
> Kinds of people – the QUICK and the DEAD!

The warden's spot reports (and reactions) formed the feed lines of ARP through the posts to Report Centres (roughly one per 100,000 of population) on to the Control Centre and the authority's controller. Those who at first volunteered for this service were another self-selected minority slice of the citizenry. A Mass-Observation study of 1000 wardens prepared well before war broke out, gave 82% as claiming 'patriotic' and sense-of-duty motives for joining, with the influence of family and friends as much the largest single factor stimulating the decision. Nevertheless, it was widely said that many had joined to shelter from military service or for some other unworthy motive. Only 400,000 wardens were paid full time; and the government started economizing on these after the war – but before the blitz – began. Many part-time volunteers dropped out early from boredom, too.

The whole population was to participate in ensuring darkness as a form of passive civil defence. The blackout, first applied on 1 September 1939 and enforced by the wardens, in one sweep changed nocturnal habits. At first this gave rise to much annoyance and a large though temporary increase in road accidents and minor injuries. Though people rapidly adjusted to the limits on light at night, these were never gladly accepted, remaining a major inconvenience in people's minds all through the blitz and until war's end.

Wardens had another important role as planned: marshalling people to and from public shelters in response to the warning sirens; long wailing cries for 'alert', warbling 'all clear'. Once the alert had sounded, the population was supposed to go either to a private shelter prepared in home or garden, or to a public one prepared through the local authority. But as with evacuation – and indeed *all* pre-war official policy – the wish was to keep civilians as dispersed as possible. This was not based on any demonstrated facts. Professor Haldane had calculated statistically (in *Nature*, October 1938) that bombing would, in this respect, be no more and no less effective whether people were dispersed or concentrated. Official thinking went back into those roots of Douhetism. Whitehall had long since decided that there must be no 'shelter mentality'. If big, safe, deep shelters were established, the people would simply live in them and do no work. Worse, such concentrations of proletarians could be breeding grounds for mass hysteria, even subversion. The answer was the Anderson shelter (originally named for its designer, Dr David A., not Sir John), with fourteen corrugated iron sheets, 8 cwt., taking 4–6 bodies in the garden (if you had one).

Very effective the Anderson proved, too, when the blitz came. But it did not deal with the non-physical concerns of many people, who looked for, needed, something more indoors. These people were to force the government to modify its stand in London, notably by opening the Underground. Until then, reinforced basements and the like were prepared by many factories and businesses, while the government erected brick structures, 'surface shelters', on pavements and wasteland, intended only to take people caught out in the streets or otherwise temporarily off base. Many of these brick shelters turned out to be improperly built, very unsafe.

The first solid indoor shelter, to protect people inside the home, was named after Herbert Morrison, as Minister of Home Security, when publicly tested by Churchill at 10 Downing Street in January 1941. Four hundred thousand were ordered. Few were delivered before the blitz ended in May. A solid rectangular steel frame, 6′ 6″ × 4′, 2′ high, it served as a table by day.

Domestic shelters and gas masks (never used) were two protective gestures from government. A third was the stirrup-pump. Designed to enable one man with a bucket of water and some sand to put out a young fire, and costing 12s. 6d. it was first allocated to warden's posts. Three weeks before war started only 8500 had been delivered to local authorities, many being diverted for private sale at higher prices. A shortage of pumps continued into 1941. As late as June 1942, 95,000 were on order, undelivered.

Organization of fire-watching, left on a loose basis in the plans, was slow to mature in war. The Fire Watchers Order of September 1940 took a small first step towards overcoming initial apathy, but 'even in the summer of 1940 no one clearly visualized the need for an army of workers to help' (O'Brien). Powers of compulsion had to be used in January 1941, by which time London had received nearly 20,000 incendiary canisters. Although this cascade had far less effect than H E upon the humans below – and therefore on the main story in the following chapters – it of course did a good part of the material damage to urban Britain, notably in 'the second Fire of London' on 29 December 1940.

Once 'established', fires became the concern of the Fire Brigades. Under the general surveillance of the F B Division in the Home Office, this system, like most others, went to war retaining virtually complete local autonomy though one authority could call upon another by emergency plan. London, as ever, was a special case. The main London Fire Brigade had 61 stations, plus 66 other, separate brigades co-ordinated in the Greater London area. Outside London, however, co-ordination had barely begun. Provincial blitzes were to over-strain the arrangements again and again, as with 500 major fires on Merseyside, 20–22 December 1940; 600 obliterating much of the warehouse and business centre of Manchester in one night. Far less thought went into planning against fire than gas or HE. Fires were not new. They happened in peace, too. A National Fire Service, painfully overdue, was not formally brought into being until August 1941, nearly the third year of war.

Most people, therefore, at first thought little if at all of fire risks. On this and much else, most civilians awaited government orders. Blackout set a pattern of compulsion which seemed to make sense. Presumably, then, anything really important to the war, like conscription, would be obligatory. The vast effort of the Ministry of Information and other agencies to ask people to do things on their own, like carrying their gas masks, fell on deaf ears. With fire-fighting, the official propaganda bordered on the inept: demonstrations were held all over the country in the summer of 1940, to stimulate public response. Here is one account of a demonstration on Streatham Common, south London, in July 1940, to give an idea of the tempo:

The first demonstration is of how to put out a bomb with hand and bucket. The bomb is lit on a board in the middle of the platform and burns with a very bright light, this causes much comment. A man comes up to it, puts sand around it, then on it, gathers it up, puts it in a bucket, and takes it off the stage. Clouds of dense smoke can be seen, and after a time the bomb, still burning fiercely in the bucket, is taken on to the stage again, and covered with sand until it is put out.

The reception to this is not good. Throughout the demonstration there was much laughing. A man turned and said 'I don't think much of that'. 'What to do with a Roman candle.' Other comments were mainly on the smoke: 'Can't see for smoke.' 'All I saw was smoke', and so on.

Second part of the demonstration was with a stirrup pump. Pearce says 'Only the leg part of the pump goes in the bucket', and raises a laugh. Woman says 'That's what we've found out.' Then he practises with the pump a little on the stage. Man – 'You don't usually have the pump nice and handy to the fire like that.'

Pearce tries to light the bomb while another man goes round sprinkling the straw with paraffin. The bomb refuses to light, and there are some laughs. Finally it goes off and Pearce lights the straw. This flares immediately, and flames about ten foot high appear. There is immediate and widespread comment. Pearce is seen crawling on his stomach with a hose and then disappears into the smoke and puts out the flames surrounding the bomb.[6]

The wardens, fire-watchers, firemen, all who made up the fourth army, eventually totalling around 2,000,000, still left some 40,000,000 detached civilians as potential bomb-fodder, the expected stuff of panic. The preparations to deal with these under stress were unprecedented, enormously difficult to work out in detail in advance. Inevitably, they tended to be grafted on to existing peacetime services, designed at the centre; applied locally. The problems to be faced were largely human; some physical (death, no electricity, no home left standing); some more emotional (fear for self, anxiety for relatives, no cigarettes). The former categories were, in general, clear-cut and easier to plan for than the latter, although peacetime organizations did not improve the chances of quick action in either case.

Thus, plans to help those humans who survived direct hits on their own homes were originally catered for by a service known as 'Relief in Kind'. This sounds kindly enough, if obscure, until one recognizes that the term was crystallized in section 17 of the Poor Law Act, 1930. A continuingly coercive attitude to the bombed-out derived from the Poor Law (and 'Means Test') mentality. The Ministry of Health even added a retired Inspector-General of Police to the 'Relief in Kind' Committee, with special responsibility for advising on 'mass control' and discipline of the potentially turbulent. A compassionate, urgent approach to those who suffered as a result of enemy attack was slow to develop; it struggled to emerge during the 1940–1 winter, but did not gain full acceptance until after the worst was over.[7]

The reasons for much seeming heartlessness included traditionalism, over-engrossment in major (holocaustic) rather than minor distress, and always the inability of the highly educated to 'think down' into the hearts and minds of the less fortunate, then commonly termed 'the working classes'. The only intelligence they then had to help bridge this gap was their own, except for the young science of psychiatry in the field of the mind.

Advised by the leading experts, government planned for the psychiatric as well as the physical worst. In 1938, a professional group of eighteen eminent savants in this field warned the Ministry of Health to set up, without further delay, a massive system to operate day and night services outside city centres, to cope with the broken spirits of their fellow countrymen. Three to one was the estimated ratio between psychiatric and physical casualties, going as high as 4,000,000 mental cases in the first six months of air war. These mighty calculations were confirmed, even elaborated, in 1939 by the Mental Health Emergency Committee, which told the minister to expect unprecedented rises in neurosis. One eminent, fashionable psychoanalyst, in 1940, extended this approach to civil psycho-capacity on the basis of 'the utter helplessness of the urbanized civilian today'.

This advice dictated the planning scene, once more putting the emphasis on extreme effects. Many thousands of physically sick patients were cleared from British hospitals when war began, to make space for the trembling hordes.[8]

It needs no Marxist to declare, 40 years later, that the class structure of pre-war, pre-welfare Britain had played a major part in determining the precautions taken for the population as a whole. The intentions were wholly honourable, fully serious. From 1939 to 1946 direct Civil Defence on the home front cost £1,026,561,000. Much more went on associated, parallel activities. Perhaps no system could have done better, under the circumstances. Here we can only tell it as it was.

In what followed, the decentralization of departmental responsibilities was crucial to the relevant services as applied locally. The way these worked in practice should emerge from chapter 5 onward. In theory, the main systems were administered at the local authority level thus[9]:

In service of:	*Ministry ultimately responsible*
The dead	Health
The physically bombed	Health
The psychologically shattered	Health
The hungry or dietetically disorganized	Food
Those without kitchens (via Mobile Canteens)	Food
Salvaging furniture, etc.	Home Security
Clearing up debris	Home Security
Early home repairs	Health
Water and sewer repairs	Health
Road repairs	Home Security & Transport
Traffic Control	Transport (thro. Chief Controller)
Communications	GPO, Transport
Information of all kinds	(see below)

Some special services were allocated to officers not part of or less intimately attached to the local authority, including:

Distress schemes and injury claims	Assistance Board Area Officer
Gas and Electricity repairs	Local companies
Labour supplies	Employment Exchange Officer
Food shops	Food Executive Officer
Unexploded bombs	Military Commander
Tobacco supplies	Board of Trade (Tobacco control representative)

The first of this second-tier proved to be the most important in many blitz situations (with striking effect in 1941 Plymouth); the last was responsible for most frequent minor anxiety and inconvenience. No arrangements were made for amusements and general amenities, which tended to be either ignored or arbitrarily restricted by the Chief Constable. The three F's – food, fags and fun – were each treated differently (or not at all).

More vaguely, existing voluntary social service bodies were traditionally treated as on their own, unco-ordinated. Information was treated rather like a voluntary service. It was left to an Emergency Committee or local information officer. In some blitzes these did not play any part, though experience was to show that accurate, factual information was the single most important aid to recovery, essential if people were to use the existing services effectively in times of crisis. It was evidently never foreseen that after an all-night bombardment – especially of a city centre – literally no one would know what to do. That was all part of under-valuing, in advance, what Norbert Weiner once called 'the human use of human beings'.

Smallish wonder that, far into 1941, a Mass-Observation report to the Ministry of Information included, among many conclusions, these general, repeatedly made points:

95% of attention is paid to material problems.
Strictly human problems, which eventually determine material facts like production, or 'morale' are underestimated or neglected
Need therefore is for central policy, establishment of 'morale' priorities, and these also related to material priorities
Hypothetical list of post-blitz priorities . . .[10]

3 The First Siren

Whoohoo go the goblins, coming back at nightfall,
Whoohoo go the witches, reaching out their hands for us,
Whoohoo goes the big bad wolf and bang go his teeth.
Are we sure we shall be the lucky ones, the princess, the youngest son,
The third pig evading the jaws? Can we afford to laugh?
They have come back, we always knew they would, after the story
ended.

NAOMI MITCHISON 1940[1]

On a fine Sunday in September 1939 the British found themselves in world war once more. Next day Sir John Anderson, after fifteen years of patient Civil Defence pioneering, was promoted first Minister of Home Security, combining the traditional Lord Privy Seal's office with control of the young ARP Department (under Wing-Commander Hodsoll) alongside the key peacetime post of Home Secretary. He now presided over the Civil Defence Committee.

The Civil Defence Committee consisted of no less than sixteen ministers, each in some significant way involved in precautionary problems. Six of these met daily with Sir John. There appear to be no written minutes of their deliberations, though their weekly reports to the full committee began to be recorded in September 1940. Not unnaturally, the complications of working down to and through a quarter of a thousand locally autonomous authorities throughout the island further helped to earn the CD boss another title: 'Minister of Some Obscurity.' The first months of war did little to clarify the borderlines between apathy, preparation and delay. But they did soften many earlier anxieties, giving a breathing space in which the fact of war and the fantasy of horrible death could be rearranged mentally.

Over much of England, including London and the South-East, the leaden radio words of Neville Chamberlain announcing there was a war at 11 a.m. that Sunday were swiftly sharpened by something more. Not Armageddon, but its whisper: the ululating air-raid sirens, in places made more exciting by inexperienced use. (At Leeds one of the sound operators got his signals mixed and gave the all-clear first.)[2]

Those were testing minutes of threat; for though, in the outcome, these mechanical sounds changed little, they pre-auditioned a long and presently lethal process, indelibly linked to bomb experience – so that a few still hear the wailing sometimes, suddenly, out of a clear sky, and tremble.

From a mine of ordinary diary descriptions for that morning one from a bank clerk in Sidcup (soon to be a soldier) comes as near as any to being 'average':

Almost immediately after the speech and announcement we heard the warning siren police-whistles. We rushed to the door to investigate our neighbours' reactions – the same as our own. We had made arrangements beforehand that in the case of a raid we were to go into my aunt's house, next door. This we did, collecting our gas-masks on our way, also taking inside a pail of thin mould in case of fire. We then put up our ARP curtains in order to prevent splintered glass from coming in. We also switched on the radio in order to hear any announcements which might be coming through. During these processes we were running outside to see if anything was to be seen or done. This continued until we were ordered inside by some warden. The rest of the household then continued with its business, to await the arrival of the raiders. I read the paper.

A young solicitor was a little shocked to see at least a dozen golfers continue to play during Chamberlain's broadcast. Beside the golf course the household (five family, three maids) reacted to the siren 'just like hearing the starter in a swimming race, if they are *not* very fond of swimming'. Within minutes all the many windows had been shut, the several dogs put in separate rooms (lest they panic and bite each other), the folk went down to the cellar taking:

Gas masks (all of us)
Cigarettes (all the men)
A box of sweets (mother-in-law)
2 bottles of beer (a maid)
Knitting (another maid)

A pile of books (ditto)
A bucket for sanitary purposes (myself)

The diarist himself, inhibited by the presence of his in-laws, nevertheless noted with surprise a lack of awkwardness 'between the maids and the rest'. This was the first time the two strata had sat together, sober, inside the house, except preparatory to prayer.

A lot of people went digging that morning. An architect's assistant in Ealing, West London, dug with his father and brother, as Chamberlain spoke. They went on after the siren, as 'there is not much risk anyway'. Mother was 'a little bit frightened', watching. A 48-year-old park keeper in Eltham, Middlesex, was in his garden preparing a hole for his newly delivered Anderson shelter. On the siren, some people ran into the public shelters. He fetched a couple of deck-chairs and put them in the hole, now 2½ ft deep. The sirens went on and on, out of control. He concluded it was a practice, thus got into an elaborate argument with a neighbour on German plane speeds and Eltham's vulnerability to bombing. At last an all-clear; three other neighbours, having heard he was ill the day before offered to finish the digging. He was delighted. 'Anyone who volunteers to dig for another is a hero of the first rank,' he added.

Down in Margate (Kent), a lady cookery demonstrator took the dog for an abbreviated run ('only round the park'), sent her sister to church, listened to Chamberlain and only just heard the warning – which came earlier on the south coast. She turned off the gas (as very many did that first day) and sat at the bottom of the stairs. After five minutes, seeing people still in the streets, she turned the gas on again, let the dog into the garden. The all-clear sounded; she misunderstood, turned the gas off again, went back to the stairs – but soon gave up.

Between Margate and London, a female civil servant at Croydon (Surrey) 'feels a bit funky for a while' after the warning, but went on washing clothes with mother. She looked up at a lovely blue sky: 'what a sin to start a war on such a beautiful day'. The next-door neighbours meanwhile crouched in a newly erected Anderson. Others stood around outside looking up.

In Ealing again, a chemist (unknown to that architect's

assistant a few streets away) registered 'a nasty inward fear when the warning went'. He tried to hide his upset from his wife but felt he was trembling. They both sat silent 'waiting for something to happen'. Later that night he was pleased when a second alert actually excited him: 'no fear this time.'

'My one fear is that I may not be able to *hide* any funk I may get into' – was a frequent reaction, here from a spinster ambulance attendant (age 24). She 'could have shot' two of the men at her North London Report Centre reminiscing about the previous war. Only a 'fairly recent bride' showed any visible signs of fear. Nearer the Thames, in Chelsea, a frustrated professional writer, recovering from rheumatic fever, found heart a-flutter, breath short, pressure in the head. He admitted (in his diary) 'emotional disturbance', went on to an acute self-observation which previewed what was to become, especially in London, common experience. 'Fortunately the body can only endure so much so I shall probably get over these reactions if I can keep my mind active. But it is humiliating to be outclassed in courage by every child. *I am a coward from my neck down* so to speak.' Far to the north a young man was out in the country with his fiancée when the Gateshead sirens sounded in the distance. He tried to call in his disobedient dog and tactfully steer towards a deep ditch without changing pace. But she noticed; so he had to explain. Result: 'consternation and tears, partly my fault'. A mother out for a Sunday walk was told of the alert by a motorist. She expected an immediate raid, then and there, so stayed by a ditch at the roadside and tore out 'handfuls of grass, thinking I would push the children down there and hide them'.

Total *sang froid* was the rare exception. One student teacher switched from calm into confusion. Unaware of war, wrapped in another sound-world inside the parlour of her Romford, Essex, home:

At eleven-fifteen, I was playing the piano in the front room, when suddenly my mother burst in, shouting: 'stop that noise!' and then flung open the windows, letting in the scream of the air-raid siren, and the scuffling noise of neighbours in a hurry. Immediately, my father assumed the role of the administrative head-of-the-house, issuing commands and advice: 'All get your gas masks! Steady, no panicking! Every man for himself! Keep in the passage.'

Away in Edinburgh, a medical student:

Down late for breakfast in time to hear that war had been declared. While standing in my dressing-gown I heard an air-raid warning. Much shouting throughout the house, chiefly as to whether it was serious or a practice, resulting in our deciding to treat it as serious. We therefore all rushed for the nearest lavatory, which made me think that there might be something in the school OTC jest: 'How would you deal with aircraft?' – 'Present arms and dig latrines.'

Some responded more forcefully. A 35-year-old housewife: 'And then the air raid warning! What a fluster. Suddenly after all those words, action and *ourselves in it*! People ran up and down the road. We and our neighbours *ran* to our gates. Alan (small son) is out playing and my neighbour Mr P. walks steadily up the road to find him. What a blood-curdling noise!' In a dither of uncertainty she went back indoors and took down the curtains: 'to maintain contact with the world.'

Not far away, a lad of 19 reported his father panicking, rushing round shutting all the windows. After the all-clear everyone except the diarist had 'a drop of whisky'.

A lady book-keeper had insisted early that Sunday that her husband should come with her to Whitehall and Westminster to see cabinet minsters and MPs. Hubby proclaimed a clear preference for going out into the country instead. 'Fancy,' she insisted, 'living all these years on the outskirts of London and never seeing anyone or anything.' Finally he agreed. The couple reached Scotland Yard just as the first war siren wailed. A lady stopped to ask them what it meant:

'That,' said my husband calmly, 'is an Air Raid Warning, madam.'

'But,' said I, in stupefied tones, 'they said they would not use it any more except in necessity.'

They continued past the mouth of Downing Street, along Whitehall, at the heart of what was then the empire. People started running. A man stopped and asked them, anxiously: didn't they know the streets were to be cleared? She replied that they were heading for Westminster Underground anyway. The diary continued: '"It's closed", said the man. And then I *was* frightened. Badly. I couldn't think where to go. There seemed no one to ask, everyone continued running in the direction of

Trafalgar Square. Motors and cyclists tore along with complete disregard of corners. A "brass hat" with three orderlies rushed (out of the War Office) into a waiting car . . .' Our couple ended up back home, one train, a tram and two buses later, to a new surprise: 'Reached mother without further stir. The wireless announced that Mr Chamberlain had left Downing Street just after the warning, cheered by the crowds. That puzzled me. All the people I saw there were running . . . and anyway I didn't see any crowds.' A full-time mass-observer was stationed at the dead end of Downing Street that same morning. He made no mention of cheering or running.

Those sirens impacted more violently again in the working-class part of Fulham, where another mass-observer was sitting, sipping in a café:

A second's silence, then everyone jumps up and rushes to the door. 'It's an air raid.'

'Already?'

Panic.

Police on bicycles ride up from police station down road wearing 'Take cover' placards on chest and back and shout 'Take cover', and 'Take cover' is echoed by people in café and streets. People in the street begin to run frantically. People in houses and shops run to door.

Church-going diarists noted much variation in reactions. One in a south coast town was shocked that the preacher 'dismissed the congregation (out of doors) at a time when the air raid – if any – would just be beginning'. At two adjacent churches in South London, one preacher sent everyone home, the next carried on, with a very meagre congregation – many everywhere stayed away to listen to the radio. Similarly a Quaker meeting in Hertfordshire broke up at the siren, another reacted to the contrary:

11.30 a.m. – air raid sirens began to shriek. The caretaker told us that we were requested to take shelter – but this the group refused to do – and remained as before – the meeting continuing. My own heart thumped wildly – and I felt a strong impulse to get home to mother – for I was anxious about her safety. The futility of such a rash act was obvious however – and so I awaited the invading planes and strained my ears for the sound of gun fire and explosions. None came.

Several millions did not know war was declared when the first siren went; the sound came as a declaration of war in its own right. 'Suddenly the hooter or siren sounded warning of a raid. My brother, who was in the bath, got out quicker than *I have ever seen him move in his life*. Air Raid Wardens rush up and down the street – in panic, it seems to me – and I have a queasy feeling in the pit of my stomach.' The fullest expression of what may not unreasonably be termed September siren shock comes from a lawyer:

Sunday 3rd Sept. I did not trouble to go out for a paper as I had so much blacking out to finish. Worked like a nigger and by 11.15 a.m. had made temporary arrangements for all rooms except spare room and dining-room which I had decided I should not be using. Very thirsty work, and I had just poured myself out a long drink. Turned on the wireless, and I was on the point of sitting down for a well earned rest, when pandemonium suddenly broke out – the wailing of hundreds of sirens like souls in torment. I was filled with an ecstasy of *exquisite and thrilling panic* – it was war then, or war and an air raid to coincide with it, or even an air raid forestalling a declaration of war.

I rushed frantically up and down the house, throwing hard-boiled eggs and pyjamas into a suitcase, dashed down to the garage, leapt into the car, and drove it out with such abandon that I buckled a wing and had to go forward again before I could extricate it – all to the accompaniment of the terrifying siren blasts. I shouted to wardens who were rushing past to enquire what it signified, thinking it might be the way of signifying war, but was told 'it's an air-raid', and immediately had visions of the wave upon wave of German bombers which we had been *told to expect* ushering in their idea of the 'lightning war'. Meanwhile I careered madly up to Holland Park Avenue till a warden forcibly directed me into Ladbroke Grove and made me take cover, which I did in a garage with a skylight and open front!

After some minutes I thought I heard the all clear, so got into the car and dashed on to the Convent. When I had almost arrived there, I was again stopped by a warden and told to take cover. I and party of civilians accordingly knocked at the door of one house but the good woman refused us admission. Whereupon the warden rushed up in an absolute frenzy of rage and nearly pulverized her, and she promptly collapsed and led us all down to a most evil-smelly basement, where we waited for quarter hour or so before the all-clear was given.

On arriving at the station I found everybody standing to their stretchers, fully equipped in anti-gas clothing, and not only ready but expecting to be sent out at any moment. I was told to rush and get a steel helmet and service respirator, then that there was *no* spare equipment, so a helmet was snatched off the first man we came to and crammed on my head, I was told to take my civilian respirator and as I had in my car topboots and mackintosh I was soon more or less suitably arrayed for the fray, and 'stood by' in my car with engine periodically ticking over, ready to dash a stretcher party to the scene of action when the call came.

Actually nothing happened and by 1 o'clock the alarm was quite over. *I then learnt for the first time* that Germany had been given a real ultimatum expiring at 11 o'clock that morning.

Lest it be thought that the above gentleman was especially nervous or excited, his record shows – and *Who's Who* confirms – that he ended up at the top of a vital war department of state, winning a high decoration for distinguished service.

That Sunday had elements of panic; but panic for moments – and for nothing except the expected, the fantasy not the fact. There was also equanimity: and everything in between. The range, the changeable aura of personal and group reaction is patent from the start, and will remain, to confuse every generalization, to the finish.

Few people started and continued with fixed responses to the sirens or any other of the new experiences. *Homo Sapiens* is so pliable, so adjustable. But there was the dog that *had* to be walked directly the alert sounded. The woman never failed in this first duty, at risk to herself and always to the fury of her husband. Another home had a seemingly normal cat who got her signals crossed. The warning siren became confused with the voice of a regular BBC broadcaster who gave vigorous physical training lessons. Whenever this lively female came on the air, the cat had hysteria. In raids it remained placid. A significant section of Britons, by the way, thought *first* of their pets, canaries (as we shall read) included.

Thirty-eight per cent of homes studied by our (then crude) methods had prepared NOTHING of or for ARP by September 1939. Quite a lot had to learn everything, siren sounds included. Indeed, the whole country had to learn: not least the national warning system, which in this case went off for a single French aeroplane arriving off schedule.

Most civilians got conditioned to false alarms from the alert system. They worked like practice jumps from a balloon, as training before dropping over an enemy-occupied country. That did not necessarily lead to inattention, a blasé stance. Nearly a year later, just before the blitz, a study in London showed nine persons in ten as fairly regularly mistaking cars changing gear for real warnings: 'My hearts stops about twenty times a day', was a common report. One observer describes an entire First Aid Post turning out during the night for a bus changing gear uphill.

The 'phoney war' lull before the storm really was just that. It helped steel the mind to survive. Few were so senseless as to discard all fear for the real future. Very few wished for the worst. By the time the bombs fell false alarms at least were no longer alarming. That, of itself, proved to be a contribution towards popular stability.

4 The First Bombs

'I'm getting a nice piece of steak for my lunch, my nerves want building.'

Harrow woman after first London bombs, 23 August 1940

'Oh, my heart has stopped.'

Stepney woman, East End, 23 August

'I feel I want to lean against something.'

City policeman, 24 August

There was no sign of holocaust during the winter of 1939–40. Summer brought the Battle of Britain, fought out for daylight air-dominance, plain for many to see in the sky. It was officially thought that the Luftwaffe was building up for massive attack on Londoners. Post-war research has shown that this was not the prime German intention, even had they possessed the necessary heavy bomber capacity. Hitler and Air-Marshal Goering were thinking in more military and territorial terms, on the ground as well as in the air. Destruction of aerodromes, factories and actual invasion remained their main aims. The rest was, at first, undecided – as German air ace General Adolf Gulland has lately reaffirmed, showing that invasion, economic blockade and the Douhet approach were all three mixed up in German General Staff thinking until July 1940, with enfeebling results on their activities.

On 16 March 1940, a bomb aimed (by day) at shipping up in Scapa Flow had killed a Scots cottager at the Bridge of Waith. 30 April, a mine-laying plane exploded off Clacton-on-Sea and killed two English civilians. 9 May came apparently deliberate small attacks on two villages near Canterbury (no casualties). Eight were hurt in a similar attack in the North Riding of Yorkshire, later in May. Yet it was nearly a full year from Chamberlain's Sabbath siren to the first deliberate assault on London.

In this long lull, the unfailing human – especially British – capacity to 'rationalize', to convert wish into belief and thus obscure pending unpleasantness, had ample scope. This was in one way favourable to the war-effort: it kept most people optimistic. In another way, it was damaging since the effect could easily be to encourage apathy, neglect of the precautions officially advised, and so on. They varied from individual to individual, affecting not only each 'woman in the street' slightly differently, but every A R P Controller, Town Clerk and bank clerk likewise. The balance of moods varied not only from person to person but day to night.

In the winter of 1939 we asked 338 working-class Londoners what they thought of their city's future. Would there be raids? About half did not then consider there would be any. Some of these were mothers already returned from official evacuation to the countryside – which they called 'there'. Such were the voices of that moment, at peace in war:

'I wouldn't be out of London now for a hundred pounds! We're so well protected here.'

'They look after us so well here, the balloons and that, they'll never get through.'

'I felt so unprotected, you know, there, like anything might happen.'

'I wouldn't have my children out there, Hitler could get through so easy.'

Compare the expectations of one who never left, putting it for many more: 'As a matter of fact, at the back of my mind I don't believe they are coming – but of course I know really they will.'

The other half of the 338 more simply feared the future. 'Yes,' said a young artisan, the raids would start 'at any minute now. Smash the place to pieces!' However, he had done *nothing* in the way of personal precautions. Very few expressed positive alarm, as compared with pre-war. On the contrary, a good few seemed almost eager to reach the experience, to get into it if not get it over and done with.[1]

We could find, by June 1940, over a third of London households with 'no steps taken of any A R P sort', except of course the compulsory blackout. The larger proportion by now taking voluntary steps had, in order of frequency: (1) papered or otherwise protected window glass; (2) arranged sand and/or earth and/or buckets for incendiary bombs; (3) got together a first-aid set; (4) put a stirrup-pump ready; (5) stored a reserve of food, mostly tins, rice, sugar; (6) cleared the attic as an anti-incendiary measure.

Asked how to handle an incendiary bomb, one in four Londoners got it completely wrong. Many more did not know clearly though they had some equipment with which to react.

So it continued.

In the eleven months before bombs fell on Churchill's 'fat cow', official preparations were able to benefit by supplying stirrup-pumps, Anderson shelters and presently an official issue of ear-plugs. The basis of organization, outlook, policy, however, remained the same. There was no major rethinking. Still, low down the ladder, in the towering Ministry of Information in Bloomsbury, a small new department, Home Intelligence, had been established to monitor the public for the first time. The first Director was Mrs Mary Adams, brilliant zoologist wife of a Tory MP, seconded from the BBC. This unit gained its strength in the blitz, as human problems accumulated. Meanwhile, Home Intelligence was grappling with some of the then barely understood aspects of civil behaviour – fears, post-Dunkirk despair, effects of censorship, news, rumour, etc. – as a first look at that most neglected aspect of total war, the human factor.[2]

As the Battle of Britain built up overhead, more stray bombs scattered across the land; for instance, 12 miles from a small Suffolk village which held a resident observer. Here as elsewhere 'what caused a stir was the sound of anti-aircraft fire'. The village postman's wife complained: 'It was terrible last night, and first thing this morning, and all through the night, though not so loud. I could hardly sleep at all.'

While the newsagent's wife was so alarmed she wouldn't let her son go to school ('because of the banging last night'). When someone opened the pub door suddenly, the landlord's wife exclaimed: 'My, you did give me a fright. I'm getting terrible. I've never been jumpy like this before. I shall have to pull myself together, won't I?'

Not that there was ever uniformity, even in the smallest community. One of a group of five women talking in the same bar announced: 'I'm taking the kids and going to stay in Derby with my Dad for the duration, and I'm going to have a good binge here on Sunday night before I go.'

And so she did, as the Luftwaffe slowly crept closer to London's semi-expectant ground. Targeting Britain's air defences all the time, on 15 August they hit at two aircraft factories in Croydon, then the city's main airport. Eight people were killed or hurt, a big step-up in civilian involvement. An observer was living in a nearby household, where Mr M, 30, had beeen a rather unpopular lodger ('offhand in manner, not friendly, given to causing trouble, tea-leaves down the sink'). That night after blackout he burst in, high with the delights of survival:

'We were in it! We were in it!' cries Mr M, rushing through the front door at about 10.30 p.m. 'We were *in* it!'

His voice rises to a sort of ecstasy, while landlady and fellow-lodgers gather round excitedly. They have been waiting for Mr M's return ever since they heard the 9 o'clock news; for Mr M is an engineer at the Croydon aircraft factory said to have been bombed. Mr and Mrs K, from the top floor, come rushing down the stairs to join the gathering.

Mr K: 'You're back! Here you are! How was it? How did you get on? Did you see it?'

Mr M: '*See* it? We were *in* it! *In* it!' (he repeats these words 'in it!' something like a dozen times in this first minute of conversation). 'We were *in* it, and I can tell you straight away, they didn't hit the aerodrome!'

Mr K: 'It said they did, on the news...'

Mr M: 'Not the aerodrome, they hit the edges of it, and the works but not the aerodrome.'

Mr K: 'Were you all right?' Mrs K: 'What happened?' Miss McM: 'Tell us what happened.'	All speaking simultaneously

Mr M: (making jerky, excited little movements with hands and head, his facial expression one of explosive joy and pride) 'Well, the first thing, it's not nearly as bad as they make out, an air raid!' (his excitement at being the bearer of this new information is evident in every tone and gesture) 'The noise *isn't* so terrific. The first thing we knew of it, there was a bang, not a terrific bang, it sounded as if one of the fellows had blown up one of the gas things or something like that. There was another and it wasn't until there had been three or four that we realized that there was a raid on and we were in the middle of it! We didn't get the siren at all, just there we were, working away, thinking no harm to anybody' – (here he laughs) – 'and thirty seconds later we were in the thick of it! I can tell you we dived under those benches!'

At midnight people had not stopped asking questions of M, sometimes for the third or fourth time. His final word: 'Honestly, you know, an air-raid isn't nearly as bad as they tell you.'

The following Friday at 3.10 a.m. 12 HE bombs fell in Ladysmith Road, Graham Road, Grant Road, Thomson Road, Christchurch Avenue and High Street, quintessential suburban rows at Wealdstone, Harrow. That morning's observer at the scene met with a phenomenon soon to be recognized as commonplace. 'Everywhere,' she reported, 'there were people looking at the damage or for it.' The atmosphere was noted – not without initial surprise – as 'one of excitement, and interest'. 'People looked cheerful, and though they *talked* about nerves and shock, they showed no signs of either.' Plenty were expecting, in fantasy, to be more frightened than 'in fact' they felt.[3]

Already, on 23 August 1940, a M-O report on overall London trends (written for Home Intelligence) noticed how: 'the rapidity of acclimatization among the greater part of the population has been remarkable.' There was always a smaller part though. And the next night it was tested.

Some German planes sent to attack Midland oil and aircraft installations the Saturday night got lost and dropped their bombs in London, the first on the East End since 1918. Stepney and Bethnal Green were struck, as well as St Giles church in Cripplegate. An observer went to a public brick surface shelter in a Stepney area too dense and gardenless for Andersons. She kept a log of sirens all that day – 8.20 a.m., 4.30 p.m., 11.15 p.m. At 3 a.m. the whole street was aroused by loud gunfire. Many were already inside the public shelters. Now, in one, a woman screams: 'They're coming! They'll bomb us! I can't stand it! Oh God I can't stand it!' The woman's husband tries to calm her; already six other women were crying out loud; others on the verge of tears. Within minutes, there were thirty-five people in the shelter, all it would hold, and far more than could find seating space. Since the day before a wooden form had been brought in, but it was quite inadequate. Then the silence was cut across again by the roar of guns. 'I can't stand it! Oh God save us! Save us!' cried one middle-aged woman.

'My God, we should live to see the day!' cried another, and an elderly man exclaimed: 'They're on us!'

More guns sounded, and an older woman let out a shriek. 'For God's sake, pull yourself together, Sally!' urged her friend.

There were no guns for a long time; presently people began to think it was time for the all-clear.

Man: 'Hasn't it gone yet? There's nothing outside.'
Woman: 'How do you know?'
Man: 'I've been.'
Woman: 'Well, you're not to. Don't go out there, Simie, they'll ... Simie! Come back at once!'

Simie went out all the same, and came back to report that there was nothing doing. They stayed, sirenless, until daylight.

One of two bombs had fallen, unexploded, in narrow Montague Street, parallel to the Whitechapel Road ('cobbles and narrow pavements, old decrepit houses' in 1940). One end of the street was barricaded. The bomb was expected to blow up at any minute. Crowds surged against the barricades. A harassed policeman, asked what was going on, replied: 'Ask Jerry.'

Meanwhile, around adjacent Bow cemetery, a huge crowd of sightseers had gathered; after midday more and more people were still coming in. One young woman in the crowd voiced a complaint heard a thousand times in the months to come:

'They're coming by taxi and by tandem, by car and by cycle, in prams and on foot, to see our little street! We're in the news – but why don't they (the BBC) mention the street by name? It's not *fair*!'

The name of the street was Southern Street, alongside the cemetery. Its windows were shattered, and at one end roofs had caved in. Demolition squads were removing tiles and broken brickwork, salvaging property from crumbled bedrooms. The crowds watched fascinated.

Older man: 'I've never heard anything like it – even in the last war it wasn't like this!'
Woman: 'Oh, I say, it's terrible! It's terrible! It's awful! I

don't know what to *do*! I haven't been able to sleep, I've got no sleep, I'm so tired! Oh dear, what shall I *do*?'

Nearby, in a working-class kitchen, four girls and their mother sat listening to the wireless that Sunday evening. Florence Desmond was performing in 'Hi Gang' (which often featured Churchill's son-in-law, comedian Vic Oliver). Suddenly:

Younger sister: 'This war's killing me!'
Elder sister: 'I'm – oh, we *could* sleep downstairs, couldn't we?'

Arrangements to sleep downstairs were made. Later, at 11.15, the siren goes, with gunfire. Three girls already in bed join the others, who have waited fully dressed. They all go to the public surface shelter. Tonight the mood there is more cheerful, with attempts at a singsong ('Roll out the Barrel', 'Abide with me', and 'Bye Bye Blackshirt'). Outside, as dawn's all-clear died away, overheard in the street:

Man (25): 'I know it sounds mad, but I've *enjoyed* this weekend more than I've ever done before!'
Woman (35): (Hysterical) 'Oh, shut up! SHUT UP!'

Thus East London's baptism by fire, a few scattered bombs and considerable nervous reaction. But, like baptism, this to some extent served as an initiation into a new code of living. Those who had shown fear were not regarded as cowardly. The terrified were already showing signs of beginning to adjust, to be assimilated.

This London attack had a wider effect. Churchill, essentially aggressive, ordered a reprisal assault on Berlin. On 28 August the first Berliner was killed by a British bomb, in breach of Hitler's promise of immunity from any such unpleasantness over Germany. Hitler responded, escalating. On the last four nights of August Merseyside was heavily attacked by Luftwaffe bombers, targeting loosely on the docks. Not yet eager to start all-out civilian bombing, the Führer really let go, at last, in a speech on 4 September to a group of Berlin social workers and hospital nurses, promising revenge by razing British cities 'to the ground', one hundred tons of bombs for every ton dropped by the RAF. Goering obeyed his chief reluctantly, but Hitler 'insisted he wanted to have London itself attacked for political reasons and also for retribution'.[4]

By daylight on a clear Saturday, 7 September 1940, his lightly armed bombers switched finally to London, taking existing fighter defences by surprise. They started great dock-fires before dark. Then 247 planes bombed indiscriminately until near dawn. London was not to have another alertless night until 3 November. By that time it had taken well over 10,000 tons of HE and as many canisters of incendiary. What official Britain had expected in fear for nearly two decades came to pass in its own different way. The *blitzkrieg* was on. How did the people of east and all London now react?

5 The London Blitz

'... the greatest target in the world.'

Churchill in the Commons, 30 July 1934

'That 'Itler! – Giving us the all-clear when we was settled. *Now* what are we going to do?'...

Stepney woman

'I think I'm going to die.'

Stepney girl, 7 September 1940

(1) Smithy Street to Oxford Street (and back)

This record begins at 8.15 p.m., 7 September – inside a street shelter at Smithy Street, Stepney, East London again. Already about 35 people have crowded in. Some are sitting on stools or deckchairs, some standing.[1]

At 8.15 p.m. a colossal crash, as if the whole street was collapsing; the shelter itself shaking. Immediately, an A R P helper, a nurse, begins singing lustily, in an attempt to drown out the noise – 'Roll out the barrel ... !' while Mrs S, wife of a dyer and cleaner, screams:

'My house! It come on my house! My house is blown to bits!'

Her daughter, 25, begins to cry: 'Oh, is it true? Is our house really down?'

There are three more tremendous crashes. Women scream and there is a drawing-together physically. Two sisters clasp one another; women huddle together. There is a feeling of breath being held: everyone waiting for more. No more. People stir, shift their positions, make themselves more comfortable.

Then, suddenly, a woman of 25 shouts at a younger girl:

'Stop leaning against that wall, you bloody fool! Like a bleeding lot of children! Get off it, you bastard ... do you hear? Come off it ... my God, we're going mad!'

People begin shouting at one another. Sophie, 30, screams at her mother:

'Oh, you get on my nerves, you do! You get on my nerves! Oh, shut up, you get on my nerves!'

Here the A R P *helper* tries once again to start some singing.

'Roll out the ...' – she begins.

'Shut your bleedin' row!' shouts a man of 50. 'We got enough noise without you ...' Outside the gunfire bursts forth again. It grows louder, and now the A R P girl begins walking up and down the shelter, between the rows of people, singing and waggling her shoulders; a fine-looking girl, tall and handsome, with a lovely husky voice:

'There's a good time a'coming, though it's ever so far away ...'

Older woman (to young girl sitting beside her), 'Why don't *you* sing?'

'I can't! I don't want to! I can't!' cries the girl. 'I can't! ... Oh God ... !'

The singer tries to get people to join in, but they won't. She gives up and sits down.

'My God,' says a young artisan, 'I want to laugh!'

Around midnight, a few people in this shelter are asleep but every time a bomb goes off, it wakes them. Several woman are crying. At each explosion there is a burst of singing from the next shelter. Two men are arguing about the whereabouts of the last bomb.

Suddenly a girl cries out:

'I wish they'd bloody well stop talking and let me sleep! They talk such rot ... such rot it is! That man, listen to him ... he's got such a horrible voice! Tell him to stop ... Tell him I said he's to stop, he's got a horrible voice ... !'

The girl's neighbour tries to calm her, urges her to try to sleep. No; she screams:

'It's no good! I'm ill! I think I'm going to die!'

By now the women with deck chairs are lying back in them wearily, rocking and groaning.

Woman of 60: 'If we ever live through this night, we have the good God above to thank for it!'

Friend: 'I don't know if there is one, or he wouldn't *let* us suffer like this.'

When the all-clear goes, about 4.30 a.m. there is a groan of relief. But as soon as the first people get outside the shelter, there are screams of horror at the sight of the damage . . . smashed windows and roofs everywhere . . . smoke streaming across the sky from the direction of the docks. People push and scramble out of the shelter doorway, and there is a wild clamour of shouting, weeping, and calling for absent relatives.

'Where . . . ? Where'd she go . . . ? Oh, 'e never shoulda . . .' shrieks a woman, incoherent with anxiety. Others sob and cry, and cling to one another. One man throws a fit; another is sick.

Later that day, in the windowless front room of one of the shattered Smithy Street houses, a young woman sits among the remains of her possessions, crying her heart out: it is her birthday.

'I'm twenty-six,' she sobs, 'I'm more than half way to thirty! I wish I was dead!'

All that day, in the bombed streets, there was a turmoil of comings and goings, taxis everywhere, as people searched for relatives, belongings, and some sort of refuge for the coming night. There was a massive exodus in two directions – eastward to Epping Forest and the open country where thousands camped that night, or westward towards the centre of London, where it was believed that the shelters were deep and safe, the bombing less severe.

The second route was to become habit for many. The long trek back and forth to the West End had begun. Evening after evening, thousands of East End 'slum' families set off, with food and bedding, to spend the night in comparative safety 'up West'. Follow the fortunes of two of these families, related neighbours, the D's and the K's, heading westward for the third full night of the blitz: two sets of parents, a grandmother and several children.

Preparations for the journey began at 1.30 p.m. The women had all gathered in Grandmother's kitchen. Mrs D wanted to wash her twins, but there wasn't any hot water. A large kettle was on the kitchen range; but the water in it was scarcely warm, so she decided to wash them in cold. Soon there was a furious quarrel raging; something about 'flannels' – difficult to tell exactly what.

Eventually blankets were tied, sandwiches packed, bags done up. They were to have set off at two, to make sure of getting a place in Dickins and Jones' basement, the big multiple store in Oxford Street. Yesterday they had been too late.

Two o'clock went by; it was decided to go by three. At quarter past three, only Granny and Mrs K were ready. They decided to set off with:

Granny: folding chair, blanket, slippers, bag of food.

Mrs K: bag of food, blanket.

After some while, they managed to get on to the crowded platform – where the rest of the party soon caught them up, with the addition by now of two more neighbouring families. Others were bound for Edgware, Golders Green, Stamford Hill. All carried bundles. There was more quarrelling; reproaches directed against Gran and Mrs K for having gone on ahead.

Then the siren went again.

Mr D (working himself up): 'Don't get excited, anyone! Stay where you are. Keep calm. Don't get excited' (This was his key phrase for the entire day).

Betty (aged 8): 'We'll be bombed!' She begins to cry.

Mrs K: 'We're going on, whatever happens! Once in the train we're safe.'

The train came in, already packed. The whole party managed to push their way on to it. At Aldgate more people got on. There was hardly room to breathe. At Mark Lane, yet more tried to push their way on; but by now it was a physical impossibility.

'All right, all right!' Mr D kept saying, '*Don't get excited!*' At intervals, from where he stood, he issued orders to the other passengers – 'Pass along there, please!'

At the Monument, a porter called out 'Cannon Street only!' so the party got out (Oxford Circus being their destination) and tried to change on to the central line. The escalator from Monument to Bank was completely packed. More and more people crowded in until it was impossible to move. Everyone was shouting and talking and wondering what was going to happen.

A porter stood up on the escalator rail, and began yelling instructions. He was a leader and kept the crowd in a good humour throughout:

'Now, listen, everyone,' he yelled, 'this escalator isn't going, but the one going up *is*. Now platforms 4 and 6 aren't working but platform 5 is. Number 5, Tottenham Court Road, and that's where you're all going! Tottenham Court Road!'

There were no protests – the crowd seemed to *like* being told where it was they were going, to feel that someone was in charge.

The jam lasted for half an hour. People were getting hotter and hotter: the porter kept talking away. Mrs D was in an agony of nerves – she thought she had lost one of her children. Mr D was equally worried, but showed it by snapping. 'Well, what do you want me to do – put her in a net, or something?'

'She'll get trampled to death,' sobbed Mrs D. 'I know she will.'

Eventually the child appeared, quite nearby in the crowd; and slowly, inch by inch, the crowd edged its way up the staircase, along the passageways, and on to the packed platform for the Central Line.

Three jam-packed trains passed before our party were able to squeeze on to one; and once there, in the stifling atmosphere and heat, tempers rose to flash-point. Cissie, the eldest child, flared up in argument with a fellow-passenger.

'Shut up, Cissie!' implored her mother.

'Why the hell should I?' retorted Cissie.

At last the party reached Oxford Circus; as they emerged into Oxford Street, they saw straight away that hundreds upon hundreds of people were already crowding round Dickins and Jones; hundreds more were making for it. The police were trying to prevent a queue forming. Such was the pressure of the crowd and their determination, that at last the police gave in, and attempted, instead, to insist that the queue formed itself in twos. This involved pushing people back and back from the store entrance. Everyone obeyed.

'Where the bloody hell do you think you're pushing me to?' one man grumbled.

'To Hyde Park, if necessary,' replied the constable.

The Dickins and Jones shelter was opened at 7.5 p.m. Only the first 700 people were let in (the siren went at 7.15). Down two flights of stairs, a huge clean basement, which looked very strong, thick pillars supporting the iron-girdered ceiling. The walls were painted white, the floor concrete and clean. Refreshments were being served; little wonder that this shelter was so sought-after.

Soon, the entire floor was covered with bodies stretched out on blankets. Some had brought chairs (only small ones allowed), pillows, cushions, eiderdowns, etc. Others sat on chairs which were already there, and on a big round table. Queues formed for refreshments – coffee, cake, buns, etc. – and girls were sent out among the crowd with trays of chocolate and ice-cream.

Two men came in and stood in the middle of the room, one playing a guitar, the other a saxophone. Then they came round with one collecting box for the Spitfire Fund and another for the cleaners in the morning.

By 10.30 p.m. hardly any of the 700 were moving about. The heavy drone of hundreds sleeping prevailed, until stilled at 5.30 a.m. when a warden with a loud-hailer shouted from the door: 'All Clear! All Clear.' Morning came as mornings must: the long trek home to another Stepney day for Mrs D and Mrs K and Granny.

Back there, death was as conspicuous as it was wholly, blessedly invisible in Oxford Street. Nearly a thousand were killed in the first three September blitz nights, though on the peacetime estimates it should have been many times that. Near the bombed Stepney cemetery a string of hearses excited great interest as the coffins were readied on 10 September. A woman sobbed: 'It might be anyone. It might be *anyone*.' 'Who is it?' asked another. 'What's the matter?' sobbed the first. 'It's not fair we should have to suffer like this! We never thought it coming. It's coming to all of us.'

'It has come, lady, it has come. It's here for us all. We're alive now, we might be dead tonight,' said an elderly man.

Woman, 50: 'Who is it?'

Man, 45: 'Chap and his wife and kid, from Bermondsey Street. Shelter fell in.'

Woman: 'My God, even a shelter's not safe.'

Another man: 'Of course they ain't. Did you see that one up Walton Street? Smashed to bits it was.'

Older woman: 'They say eight hundred people were killed in one, Sunday.'

Older man: 'It's bloody awful.'

Another group:

Man: 'Who is it?'

Woman: (weeping) 'They say it's a woman and her husband and three-month-old baby. They got killed in their Anderson. It's happening all round. Look up there – smoke. Look down there – glass. It's everywhere the same, there's no escape.'

'We're going up West tonight, mate.'

'What's the use – they're bombing the West now. It's the same everywhere ...'

A child of ten ran across the road towards the cars as they drove off; ran back and announced:

'It's a woman and a man and their four-year-old girl – they got killed while they were in their shelter.'

Her father: 'They say they're having another evacuation. *Now* will you go?'

Child: 'No – I want to stay here.'

Father: 'You've just seen – you want to be safe, don't you?'

Child (sobbing): 'I want to stay here with you ...'[2]

(2) Sticking it!

By the middle of September, half the population of Stepney was gone. Notes fixed on battered front doors gave new addresses: in Becontree, Chadwell Heath, Dagenham, in Stratford and East Ham – further out, but still 'in London'. Few had left altogether. In one street, only a fifth of the householders were now there by night, most by day. In less congested parts of London, few left. The pressure outlets became West End shelters and soon the huge shelters and the Underground. Those who stayed totally 'put' tended to be of 'tougher' calibre. In any case, by mid-month little remained of the earlier screaming and near-panic.

Instead, a sturdy, cheeky sort of attitude was growing. In the

Smithy Street shelter, scene of so much earlier semi-hysteria, on 19 September:

Young woman: 'Was that a gun? Did you hear a bomb just then?'
Young man (Jewish): 'No, that was somebody dropping potato ludkies.'
Young woman: 'That's funny: *we* say someone's dropping peanuts!'

The women in particular were growing tougher; even crying was becoming rare. After a really bad raid (18 September) a young woman was able to describe to her neighbours the terrible scene in Exmouth Street, with limbs of children protruding from among the bricks, etc. – and then add: 'Everyone was frightened, but they controlled themselves. Hetty cried, but she cried quietly, and no one saw her. Gertie fainted but she fainted at the back and didn't make a fuss . . . I'm glad my parents aren't here. My mother would have been hysterical.' Later that same night a warden described behaviour in his shelter: 'They got over it. A dreadful shock, and they got over it quickly. They kept bringing people in from Exmouth Street, though, and it got *very* hot . . .' This new toughness was not, of course, due solely to the disappearance of the timid – the people who had chosen to move rather than stay. It was due also to a process taking place all over London: the slow and steady acclimatization of the population to life under bombardment. We shall have much more to say later on the difference between sticking and shifting. In the provincial blitz it became of prime importance. In Greater London the issue was confused by the sheer size of the metropolis and abundance of alternatives. Basic to the whole business of living through the blitz, was the process of personal and family adjustment.

(3) 'Nippy': evolution of seven waitresses

By 9 September, while the D's and K's (like many thousands more) had started a mighty trek westward in search of safety, the bombs, too, had begun to move in that direction. Somerset House was bombed, Victoria Station, and parts of Holborn –

which, by November, had received an estimated 19.7 HE bombs per acre, more than double Stepney's long-term quota (8.3). Other districts to receive over 10 per cent were Shoreditch (16.7) in the east, the City in the middle; Chelsea and Westminster to the west in the same period. (By May, blitz-end, Holborn had had some 40 tons of HE per acre, the City 30, Westminster 29, Shoreditch 24, Stepney 20 and Chelsea 19.)

On 9 September, then, seven newly-recruited waitresses (including a mass-observer) are eating their first dinner in the staff canteen of Lyons Corner House, Marble Arch, past Dickins & Jones at the west end of Oxford Street and one of the largest restaurants in London, with a staff of nearly a thousand. These 'Nippy' trainees – as Lyons called them – are sitting together at the new girls' table in a babel of talk; 'air-raids' almost indistinguishable from 'customers', so new, so exciting and so alarming are both these hazards in the girls' lives:

'... there's not a single street in Bow that hasn't had one. We never thought we'd live through it! ... I've had one customer complain already!'

'What! Have you served a customer?' – (heads crane forward eagerly, for most of the girls have spent this, their first morning, in training classes and practices).

'Yes! (proudly) He said there was a maggot in the salad!'

'Lord Haw-Haw says he's going to bomb all the Corner Houses next Sunday because they're Jewish!' (Much laughter).

'... I was in Chelsea when the Screaming Bomb came down. It blew me right across the road!'

'Battersea got it terrible ... I was caught in a bus. I've never run home so fast in my life!'

'Never to run, she said. It gives the customer a bad impression, running. Walk briskly, she said, but not to run ... never mind what a rush you're in.'

'The customers are nicer in Bryanston upstairs café, she says, more easier. *Customers*! My, I'll be too scared to speak!'

'What with the air raids and starting off here – fancy having to start off like this with no sleep!'

All these remarks are made with great cheerfulness, and the conversation is interspersed with frequent laughter.

As soon as they finish dinner, these seven girls are told by the supervisor that this afternoon, they are due to go to 'school' at the Strand Corner House, Trafalgar Square, where a special course is being held for new recruits.

There is much talk and laughter as they stand waiting at the bus stop. Loud comments are made about passers-by; two Belgians in plus-fours cause great merriment. The bus-conductor is drawn into the noisy joking:

'Do you know, you mustn't talk at night now when you're in bed?' he says.
'Why not?'
'Because there may be a Jerry under the bed!'

Yells of laughter. Other occupants of the bus keep their heads averted. Arrived at Strand Corner House, the seven from Marble Arch are joined by new employees from other branches, making a total of 16 girls and 9 men attending class.

The instructress is a woman of about 40, with a pleasant, rather worried face. She begins with a general talk about the job: the keynote of the Lyons service is that it is absolutely standardized, so that customers in all branches will meet with the same service and manner: 'We don't standardize your faces, because we can't; but we standardize everything else! ...' and 'You have to remember, while you are here you are a cog in a vast machine. You must think of yourself as a cog in the machine, and forget yourself.' As the class went on and evening approached, edginess shows among the girls; they jump and start at sudden noises – boxes dropped overhead, cars starting up outside. At about 4.30 (when the class has been going on for over an hour) there is a whistling sound from outside. One of the girls cries out, and two or three start from their seats. The instructress frowns at them, and continues her lecture ... but before she has completed so much as a sentence, the siren sounds.

There was a sigh, and 'Oh's' and 'Ah's' of dismay, tinged with amusement.

'There – you tempted Providence!' says the instructress to the girl who had first cried out; and she goes on to say that the instructions are that there is no need to go down to the shelter until the 'danger overhead' signal is sounded. 'We have a watcher on the roof,' she explains.

The lesson continues: how to change a cloth on a table – a very elaborate process, the object seeming to be never to leave any of the table uncovered, as if it was a hospital patient. One man gets completely tied up in the two cloths: everyone is laughing.

Just then the Danger Alert is sounded – three sharp blasts on a hooter . . . 'They're overhead – we'll have to go down now,' says the instructress. Everyone gets up, rather slowly, and begins to file out of the room. There is much backchat and laughter as the girls pass the politely waiting bunch of men by the door. The men display a jeering attitude to the whole thing: the girls, though they giggle, show signs of trepidation; almost all are holding hands.

The shelter, deep in the basement, consists of a vast stone passage alongside the boiler room. Along the walls are great pipes, with loud sounds of machinery; benches here and there along the walls.

Soon the place is full, 300 or more employees, most of them standing or leaning against walls and pillars. A buzz of talking, mostly speculations as to how long the warning will last? – is it the beginning of an all-night raid? – and if so, what shall we do?

After three or four minutes the instructress appears: 'It seems to be quite quiet, shall we go up again?'

Lecture resumed. At the end, the whole class goes to the mess-room for tea. Still eating, there is the unmistakable sound of gunfire. Two of the kitchen lads run to the window and open it with exclamations. One of the senior men yells at the top of his voice, on a note of great urgency:

'Shut those windows!'

The boys do so; there is something of a commotion. Girls get up from their chairs: 'What's happened? – What is it? – Is it a bomb?' Someone draws the blinds. One of the kitchen porters growls:

'Sit down! What's the matter with you? Such a bloomin' fuss . . .'

Tension relaxes. Many of the girls do sit down. Gunfire continues, though it is not easy to distinguish it from the general roar of noise of the work going on in the building.

Firm's warden comes in.

'Clear the mess-room!' he yells in dramatic tones; and for the first time there is a real sense of panic. The female kitchen staff are thrown into a flurry, shouting, pushing, rushing aimlessly about. Further yelling by the warden produces some order. Once again the procession files down to the basement, this time much more speedily – no one stopping to wait for anyone else.

Once settled in the basement, a degree of calm is restored. Some of the girls begin knitting; there are complaints of the heat, and a certain amount of joking:

'If I could get at old Hitler,' says one of the girls, 'I'd cut him in slices very very slowly and cook him in a pie!'

'Serve it on the menu – 'Itler Chantilly!' (reference to instruction in the afternoon's class) – and a general laugh.

A third girl begins to chant:

'*Old* Ribbon-drop, *old* Ribbon-drop,
Cut him up in little bits and *put* him in the pot!'

After a quarter of an hour, word passes round that danger is over, though the warning is still on. The warden announces that anyone who wishes may stay down here, but that the building as a whole is now carrying on. Back to work.

This is Monday. Each succeeding afternoon a siren sounds during the afternoon class, followed, with monotonous regularity, by 'danger overhead'.

During this first week of bombing, there is a very good atmosphere prevailing. Relations between management and staff are cheerful and friendly, and great consideration is shown towards the employees' problem in getting to work. Late-comers are not reproved; the supervisors make a point of chatting with the girls sympathetically about the previous night's experiences, and asking them if they feel all right. The Uniforms Supervisor, chatting with girls in the cloakroom on their second morning, tells them in a most friendly way which of the printed rules they now 'wink at' – e.g. the rule that white collars must be *sewn* into dresses, not pinned. She sympathizes with how sleepy they must be feeling, tells them not to worry about being exactly on time in the morning. The Floor Manager repeats this assurance.

'We're winking at five minutes, these days,' he says.

The girls (two from the East End) seem to be in good spirits while on the job. Last night's experiences of bombing are treated more as stimulating food for conversation than as reasons for depression.

The fourth alert, 11 September, comes again in afternoon class. As the class herds out ('Ah-ee – no peace for the wicked') men and girls now push through the doorway together, with giggling and backchat, without any sense of impending danger: the thing has already happened that often.

Down in the basement passage the uproar is continuous. Several hundred employees congregate, talk and laughter mingles with the noise from kitchen machinery and the electric plant, an indescribable din.

By Friday, however, the novelty is wearing thin. In the cloakroom of the Marble Arch establishment, it becomes clear that afternoon that several girls have no intention of making the customary journey to the Strand. 'We'll get a raid, and have to stay there all night. I wouldn't live through the night if I had to stay there – I wouldn't, honest! Down there among those horrible pipes, it's a deathtrap!' An alternative plan is mooted – of going to look at the damage to Buckingham Palace, bombed for the second time that day. But news of a bomb landing 'just across the road' shakes the party. In the end, three opt for going to the Strand as usual. Only 8 of the 25 arrive at class. The rest go home.

This large-scale truancy on Friday afternoon marks the beginning of a downward trend. From then on, attitudes to the job become steadily more and more slapdash. Not only the recruits but the established waitresses as well begin coming in later and later – sometimes three hours behind their scheduled time. At the beginning of the raids, lateness has met with tolerance; by the second week, the Lyons authorities are fed up. 'Well, it's the same for everybody,' they retort as yet another girl launches into her bomb-story of the previous night.

There is almost a competition in the mornings to see who can come in latest, and get away with it: 'I've been late every day this week, and I didn't come in Thursday at all! I told her, I don't care!'

Attempts to check this landslide of unpunctuality are futile, and often met with open defiance: 'I told him I wouldn't work after half past five: I just said I wouldn't. I'm not going to do it, I said, and I haven't!' The supervisors, once well-liked, are becoming more and more unpopular. Complaints about the work are by now non-stop. Tips are falling off badly, with the blame for this distributed indiscriminately between customers and management.

By 19 September, all vestige of 'loyalty' to the firm seems to have vanished. The previous night a bomb has dropped just beside the Corner House, damaging the gas-mains; and there is an unexploded bomb in the building itself. Part of the restaurant has been roped off. There is a chorus of protest going on – not about the danger of the unexploded bomb (the presence of police and wardens give a sense of security there) – but about change of work to a different floor, the Quebec Restaurant.

Quebec floor, when the girls reach it, is in a turmoil. Unoccupied staff are milling around everywhere. Two supervisors are standing distraught among crowds of girls, talking, jostling, asking what they are supposed to do?

At about ten o'clock a supervisor mounts the platform and makes an announcement:

'We have decided,' he says, 'to open this café to the public. I want you to listen to me carefully ...' various instructions ... 'Nothing is "on" except what is printed on this slip. Listen carefully and I will read it to you.' He reads list of items: the only hot ones are soup, steak-and-kidney pie, and Brunch. 'All salads are on, and, of course, tea and coffee' – at this there is a general laugh, which seems to surprise him. 'Now, girls, this is an emergency, and we are relying on you to do all you can to help us ...'

His appeal falls on deaf ears. As soon as he has finished speaking, an angry chorus:

'Isn't it ridiculous? We'll be doing nothing all day. Might just as well let us go home and get some sleep ...!'

'I'm going!'

'I'm not asking. I'm going!'

By now it is nearly lunch time (11.15 a.m.) and more girls come up to the mess-room. A notice had been put up on the hatch: 'No gas for cooking. Please help by taking only one dinner per person.'

The effect is zero. The same number as usual (about 40%) take second helpings; the question is, not 'Ought I to take more?' but 'Will I get told off if they notice?'

This lack of co-operation continues throughout the following days of upheaval due to bombing: the difficulties and disorganization simply provide that much more opportunity for getting away with skimped work and unpunctuality. On the evening of 23 September, the king speaks to the nation from Buckingham Palace 'with its honourable scars'. A weary manager at the Corner House sums up *his* vision of things so far: 'Morale here – ? It just simply *isn't*! The place is all to pieces . . . Nobody turns up on time . . . we don't even tell 'em off any more, what's the use? War Effort, my foot! Hitler can walk in tomorrow, for all the effort that's going on here!'

Such work dislocation was among the targeted effects of civilian bombing, of course. It was, up to a point, an inevitable element of the total dislocation. The record shows that where, as at Marble Arch, leadership was diffuse, the damage was greater. Industrial and commercial administration was subject both to the same strains on the spot and to the same defects in anticipation as the official machine, though normally its inadequacies affected less people. Some concerns showed even more dislocation, others much less. The factors involved were numerous, the evidence collected at the time exceedingly meagre. But the overall effect of bombing on work patterns was heavily in the direction of minor dislocation and delay, rarely towards anything approaching breakdown – as will be seen in more industrial contexts, in later chapters. Nor was there ever any suggestion that Lyons service, either at Marble Arch or in the Strand, broke down. The customers went on being fed, only more erratically and with less variety. 'Morale', as the supervisor saw it, had gone to pot because full discipline could not be enforced. Perhaps he was asking for miracles? His concern was with only the daytime part of the story: the other started at home, at night in Stepney, Holborn or Hackney Wick.

'High' morale at work need not mean the same thing as 'high' morale in a raid. One of the surest ways of being upset during a raid turned out to be thinking too much about it. 'High' morale then could consist of ignoring it – as more and more people learned to do, in various degrees. 'High' work morale nearly always consisted of concentrating on the job.

When the blitz burst upon London it brought an entirely new element into people's lives, one for which they had to find not only the courage and the stamina, but also the *time*. An air-raid – especially if you take all the appropriate precautions – can be an extremely time-consuming thing; to the novice, it is like a second, and very demanding, job. Thus, working people were confronted with the primary task of combining their earning-lives, their air-raid lives and their personal lives into some sort of viable whole.

For the first few days of the London blitz, social life was shocked almost to a standstill: one left work in the evening to go home to the air-raid. One emerged from the air-raid in the morning to go back to work, maybe late, that was all; and while it was new, exciting, overwhelming, it was enough. Few had the time or the emotional energy for anything else – at first. Gradually, as the nights went by, priorities began to shift. Home life began to acquire some pattern again. New infrastructures were evolved, suited to the new conditions. Routines were established – going to the shelter or not going to the shelter; eating early before the sirens or packing up a picnic. The repetition of bombing on London, every night, helped give such routines both urgency and rhythm.

So, too, the rhythm at Lyons changes tempo in late September.

Round the staff entrance they are gathering: the boy-friends, unseen and almost untalked-of for days, are on duty again by the third week, hanging about to take the girls out. Anecdotes at meal-times are no longer primarily bomb: more and more of them are liable to be like this (22 September):

1st Waitress: 'Didn't you go out with him, then?'

2nd Waitress: 'Not with him, no, with my friend. She's engaged, see, but I told her, I said, Don't you tell me off if I find a fellow – you know, kind of thing. We were right by Marble Arch and two Air Force fellows came along, and they kind of looked at us – you know – and we kind of looked at them, kind of thing . . .'

She goes on to describe an evening's drinking, going from pub to pub.

... 'We ended up, they took us down to Victoria we went to a milk bar. And when we come out of there, the guns was going off and the bombs dropping, it was terrible.' The possibility of going down to a shelter and the four of them spending the night there was discussed, but the girls had been afraid to agree because of the trouble it would cause at home – 'We wouldn't half get ticked-off!' The party had set off home through the bombing. 'It was awful, bombs dropping all the way along. When we got there the fellow with me wanted to kiss me good night. It's a cheek, don't you think, when a fellow's only known you half an hour ...?'

This anecdote is listened to with great interest by a whole table of girls.

Many are the complaints – related with a certain pride – of spoil-sport parents who try to get their daughters to come straight home and shelter with the family. Six Nippy voices:

'We didn't half get ticked-off when we got back last night! We knocked it up we'd been having a meal. Well, she didn't half tick us off!'
'Expect us to stay in all evening for an air-raid? I'm not going to! It's awful, isn't it? – Stop in every evening!'
'They've had *their* time, and now they expect us to stop in every flippin' evening! Some hope.'
'... down the shelter the minute you get in from work, that's no life, is it?'
'Oh, blow the bombs is what I say. I mean, you can't expect a feller to stay with you if you never go out with him of an evening, can you?'
'We're only going to the pictures, I told her, it's safer at the pictures than what it is here.'

The bomb as background is becoming part of the every night scene. These six Nippy girls, persons with little education and no abnormalities of note, have separately and collectively adjusted, managing to resume the main interest of their young lives – a job of sorts, boy-friends, cinemas and pubs, West End outings. They do so in the teeth of Luftwaffe, Lyons, the landmines, Douhetism, blackout, fear of death, the lot. Who ever imagined *that*, in Whitehall, in 1938?

(4) That 'my bomb' syndrome

One seemingly essential piece of the adjustment was to experience a bomb closely, for yourself. In London this soon became easy, so indiscriminately grew the scatter. This personal adventure was most readily achieved, however, in your garden shelter or home where it was individual in a way more difficult inside a communal shelter of some kind, with many sharing in the incident.

Associated with direct experience was the general atmosphere of excitement, adventure, a form of 'heroism' by escape, which we already noticed momentarily in the first bombs of August. Sharing in such drama was, by the standards of everyday living then, a terrific experience in its own right. The effect was, in more cases than not, to ease the burden of previous anxiety or present fear. As we had put in a report to Home Intelligence before the blitz got really under way: 'In the immediate area of a severe raid, there is a period of shock and mental blackness. Usually, rapid readjustment follows, and most people are outwardly back to normal by the next day, though still getting special liberation by recounting their own experiences with intense excitement, in purely personal terms. Many get a great deal of pleasure from having been in the middle, it makes them feel braver.' Once the blitz came, bombs all about, Londoners almost invariably (we found) erred on the side of thinking the heard bomb was nearer than it really was. 'We're in the front line! Me own home – it's in the Front Line,' declared a grizzled, elderly Cockney, staring at a wrecked house (not his own). People came from far away, at first, to look at the nearest damage. Sightseers from more than half a mile:

'We thought it was just out the back, you know. I could have sworn it fell absolutely in the back garden here. My, when I heard that whistling I thought it had Number 60 on it!' (laughs)

'When the explosion went off everybody (in his house) thought it was right here.'

'When I heard that awful whistling, when it began, I went all sort of numb. This is it, I thought; this is what it feels like when it gets you, all sort of numb. It seemed to be coming right at us, dead straight, right at this house ... I couldn't credit it this morning when they told me it had fallen on the Grange (Cinema, half a mile away). I just couldn't credit it.'

Edward Glover, Director of the London Clinic of Psychoanalysis, claimed in a broadcast at about this time that this feeling of the bomb 'coming right at me' was peculiar to 'over-anxious superstitious people who imagine that every bomb is aimed personally at them'. There is no factual support for this Harley Street view. People of all sorts and temperaments, of all degrees of 'courage' and 'cowardice', were liable to describe their first experience in these terms.

A tall young woman, who had been walking with her boyfriend in South London the night of 7 September at 8.30 p.m., told her diary next morning:

When the air raid siren sounded at about 8.30 p.m. I was walking on Wimbledon Common with C. We had been expecting the siren to sound at any moment, thinking that the enemy would of course take advantage of the (daytime) fire raging in London. Searchlights flashed around the sky, ack ack fire could be heard and the drone of a plane high in the sky. We sat and watched for a little while, but I disliked being still, and we walked homewards via Roehampton Estate. Had been walking across the common for about five minutes, my greatest desire being to keep under the trees, and came out on to Telegraph Road. We were walking along the centre of the road, although all the time I was eager to walk to one side under the trees. Four young boys passed us, going in the opposite direction. They seemed quite unperturbed. We could still hear a plane and see the searchlights trying to locate it. Suddenly a swishing noise came creeping along to the left of us, increasing in force and sound as it came. The noise seemed parallel to the ground. C threw me on to the ground and covered my head. As I fell, triangular spurts of flame seemed all around; the whole effect was like a gigantic jumping cracker. I seemed quite numb, my mouth was full of grit and dirt. All was quiet, and pieces of earth showered on to us. We both got up. Laughed a little – we were both terribly dirty. Clouds of smoke were all around and a very strong smell of earth. We both walked down the road towards home. I felt very excited, flushed and warm. Ready to laugh a lot, and probably talking a great deal...

Two days later, on 9 September, Hampstead in North London got its first bombs. A young woman happened to be staying the night in a house a few doors away from a direct hit. She, too, was in the company of her boy-friend, and, in this case, of his parents as well, Mr and Mrs R. Next day she wrote at length; here is part of it:

Mrs R made us all some tea – just for something to do, I think, because actually we were all awash with tea already. That's one trouble about the raids, people do nothing but make tea and expect you to drink it.

There was a nasty, nervous feeling in the kitchen – I don't usually feel nervous when E is there, but I did tonight, and so did everyone. Every time there was a thud, Mrs R would say, 'Is that a bomb?' – and Mr R would say, 'No, it's a gun, dear.' As time went on, they'd said it so often that it got like this: Mrs R would say, 'Is that a bomb?' and Mr R would answer 'No, s'a gun,': and then just 's a gun,': and then finally a noise that sounded like 'Sgum'. She could see he was getting irritated, and tried not to do it; muttering the question under her breath instead; which of course was more irritating still. I felt all swollen up with irritation, a bloated sort of feeling, but actually it was fear, I knew very well. A horrid, sick sort of fear, it's quite different from worry, much more physical. But I wouldn't take it out on Mrs R, I swallowed my irritation, she was feeling bad enough already.

E says to me, 'Let's get out of here, let's get some air,' and out we go, into the garden. It is a beautiful summer night, so warm it was incredible, and made more beautiful than ever by the red glow from the East, where the docks were burning. We stood and stared for a minute, and I tried to fix the scene in my mind, because one day this will be history, and I shall be one of those who actually saw it.

I wasn't frightened any more, it was amazing; maybe it's because of being out in the open, you feel more in control when you can see what's happening. The searchlights were beautiful, it's like watching the end of the world as they swoop from one end of the sky to the other. We didn't see any planes, though we could hear them bumbling around somewhere.

We sat on the grass. It is very long and unkept – Mr R is normally a besotted gardener, but he has let the garden go to rack and ruin this summer, I suppose the uncertainty and everything.

Two bombs fell in the distance. As yet she felt no fear. Her boy-friend said he could *smell* burning from the docks. Nonsense, she scolded. They talked about his pending call-up and her father (whom he disliked). Then, action:

Another bomb, nearer this time – and then, suddenly, the weirdest sort of scratching sound just above the roofs – the sound was as if someone was *scratching* the sky *with a broken finger nail*. It lasted a second, no more, and then there was the most God awful crash – it seemed only a couple of gardens away (actually it was two streets away, in Upper Park Road), I felt the earth juddering under me as I sat.

'Hey – Not too healthy out here!' exclaimed E, and dragged me up from the grass – lucky he did, because I didn't seem to be doing *anything* – just sort of cowering, I don't know what I meant to do, or what I was feeling. It was a funny, blank sort of moment.

I remember racing towards the house, E pulling me, and yelling something. There was the oddest feeling in the air, all around, it was as if the whole air was falling apart, quite silently . . . and then, suddenly, I was on my face; just inside the kitchen door. There seemed to be waves buffeting me, one after another, like bathing in a rough sea. I remember clutching at the floor – the carpet, or something – to prevent myself being swept away. That's how it seemed – this smell of carpet in my nose, and trying not to be swept away, and I could hear Mrs R screaming. E was nowhere, the lights were gone, it was all dust, I didn't even wonder if he was all right. It was funny, that, because I worry a lot about him in raids normally. Anyway, this time I didn't give him a thought, seemed to be nothing in the universe but this dusty carpet I was breathing, and having to hang on like grim death. I clutched the floor as if it was a cliff-face that I had to hang on to, or else I'd fall – actually, I was lying down already, horizontally: why I had this feeling of saving myself from falling, I don't know. It was very strong. Mrs R seemed to have been crying for ages, and calling out something or other, but somehow I didn't think of answering her. Actually I don't know if I could have, because I discovered later that my mouth was full of plaster and dust, but I had no feeling of this at the time. I just didn't *think* of answering, or of doing anything about anything. It was almost a tranquil feeling. I could hear Mr R yelling: '*Down* everybody, get *down*. Get your heads down! Do what I tell you, get your *heads* down!'

Over and over again he said it – no sense in it, because we'd *had* the bomb now, and everyone *was* down, heads and all, just about as down as they could be!

E had a torch by this time, he flashed it round a bit. There was plaster and glass everywhere, mountains of it. The whole ceiling had come down, it looked like a building-yard, bricks here and there as well. You couldn't see the furniture, only the clock, and a cushion sticking out.

Mr R was still shouting, giving orders, very contradictory: 'Don't move! Stay where you are! Get the H's (next door neighbours). Someone tell the H's. Don't move till I tell you...!' and so on and so on.

We got to the front door, quite a job, as the ceilings were down right along the passage, and pitch dark. I could hear Mrs R stumbling along behind us, and Mr R scolding her, on and on: 'Watch out ... Can' you look where you're going?' every time she stumbled. Actually, he was stumbling just as much, and how could she 'look where she was going' when it was pitch dark? (Mr R is normally a very amiable, detached, good-natured sort of man who rarely raises his voice).

We got to the front door, and it was wedged tight, it wouldn't budge. There were men shouting outside, and someone saying in a high-pitched voice 'Naomi, ask Naomi, Naomi will know. Has anyone seen Naomi?...'

We went back through the front room, it wasn't so bad there, the ceiling had held, and we crawled out through the broken window. It looked so bright outside I couldn't believe it, a sort of white haze, a halo over everything, though there was no moon. Two of the women cried when they saw us, how terrible we must have looked, smothered in white plaster all over, and streaks of blood from the glass. 'Are you hurt...'... 'Are you all right?'... people kept asking; and it was only then – silly though it sounds – that it occurred to me that I *might* have been hurt! That I had been in *actual danger*, really! Somehow, right up to that minute I had taken everything for granted, in a queer, brainless way, as if it was all perfectly ordinary. I never even *thought* about injury or death, for any of us. Actually, we easily could have been dead, the roof was in danger of collapse, and no one was allowed back in.

We spent the night at the F's (neighbouring street), Mrs F insisted on piling about seven blankets on top of me, and a hot-water bottle as well. 'For the shock,' she said; and when I pointed out that I was feeling perfectly all right, she referred darkly to '*delayed* shock' instead, implying that this dread phenomenon would surely hit me before the night was out.

It didn't, though. I lay there feeling indescribably happy and triumphant. 'I've been *bombed*!' I kept on saying to myself, over and over again – trying the phrase on, like a new dress, to see how it fitted. 'I've been *bombed*!... '*I've* been bombed – *me*!'

It seems a terrible thing to say, when many people must have been killed and injured last night; but never in my whole life have I ever experienced such *pure and flawless happiness*.[3]

Quite different in tempo was the experience of an ARP warden not far away in North London. At 11 p.m. on 17 September:

I was on duty at the post with Mr K (age 35) when the explosions occurred. They had just come in from patrolling, and Mr K had expressed the opinion that 'It was going to be hot tonight'. The gunfire had been pretty heavy for some time, and there had been several explosions down Kilburn and in Marylebone across the road.

There was a sudden rushing sound – not a whistle – and Mr K just had time to exclaim 'That's close!' – when three loud explosions shook the room, and we heard some screams from a neighbouring house. We rush out into the road, and find ourselves on the edge of a dense cloud of dust. One could feel the separate particles tingling against one's face, and it was difficult to breathe or see anything. A man (about 30) rushes from the other side of the road yelling to us:

'You can't go down there, there's a bomb. You can't go down there. There are bombs. You can't go down there!'

We push past him and plunge into the cloud of dust. It is a lovely moonlight night, but here it is impossible to see. A strong smell of coal-gas directs us across Abbey Road to the lower half of Boundary Road; and after about 40 yards we find ourselves on the edge of a crater about 15 feet across, exactly in the middle of the road. Boulders of clay have been thrown up all round, and there is a rushing sound of water and a greatly intensified smell of coal-gas. Mr K has the report forms, and begins filling one up *in the dark*, while I stumble about trying to find out if there are casualties.

Sent back to the post to report, after sundry further adventures, our warden returned to the crater and continued:

The wardens all seem completely oblivious of the gunfire, which is definitely heavy – quite as heavy as it is on those occasions when wardens mostly refuse to patrol because of shrapnel. Now they are standing about quite aimlessly, several with helmets off, in the middle of the road. There is no reason why more than two should remain, but actually eight do so. There is an atmosphere of great excitement and jocularity, men climbing about over the lumps of clay, talking about where they were and what they were doing when the explosions occurred.

Mr K comes up from the third bomb, which had fallen on a studio flat in Greville Place, about a hundred yards away. He is in a state of great excitement.

'The crater down there is terrific,' he says. 'I've just crawled in to have a look – it's right at the back. Bricks all over the road ... Come and look ...'

I go with him to Greville Place. The dust has almost completely cleared now, and the full moon lights up the scene almost like day. A smallish house has been demolished, and lies in a great heap of white bricks. Floor boards and bricks are strewn all over the road for many yards. In the middle of the heap the remains of the roof is caving in; it is under this that Mr K had crawled to look at the crater at the back.

K invited his colleague to join him in further subterranean explorations. In they went again:

We reach the far side of the ruins, and see the crater brilliantly lit by the moonlight. The studio had contained a lot of statues, and among the heaps of brick one would suddenly see a white hand sticking up in the moonlight, or a piece of a trunk, or a face. The effect was uncanny. We look round in silence for a moment. We climb about exploring (quite irresponsibly – we know there are no casualties) – experiencing just the same sense of unreasoning exhilaration that is obviously affecting K. We call to each other about the various oddments we come across – bits of books, a bedroom slipper, an amazingly undamaged clock.

There is a sudden rustle, and we both stop dead.

'Anyone there?' calls Mr K.

We satisfy ourselves that there is no one there, and crawl back. There are three or four wardens and police outside, all in a very school-boyish mood, teasing and back-answering. Mr K relates our adventures to them all as we walk on in a body to the scene of the third bomb (garden of a block of flats).

We climb about like school-children. An aeroplane is heard throbbing heavily above.

'Come on, Jerry, come and have a look!'

'Poor old Jerry! Thought he'd got the railway, didn't he?'

'Never mind, Jerry. Better luck next time!'

The aeroplane moves off to the north.

'Cheerio Jerry! See you tomorrow!'

When, still in this state of exhilaration, the adventurers returned to their post once more, they noticed an 'immense contrast' between those who had had to stay down in the dug-out (on telephone duty or as reinforcements) and those who had been out on the 'incident'. The latter were in a high state of exhilaration, and appeared to be experiencing a sense of great physical well-being; the former, on the other hand, looked bleary-eyed and miserable, half dropping with sleep, and quarrelsome. 'They stare at us and our *hilarity* as if we were beings from another world.'

'I'm feeling like another bomb now, aren't you, boys and girls?' said the Deputy Post-Warden (who had been out all the time).

'What a thing to say!' said Mrs L (on telephone duty), in a cracked, tired voice. 'I'm feeling just about *dead*.'

There is no need to elaborate the dynamic of this experience as recorded on the diary seismographs. Let us, rather, change emphasis and listen to five men taken *talking* at random in another part of London (Mill Hill) after the first severe raid there, towards the end of September, as reported next day by a resident observer who recorded a general attitude of 'excitement', while ARP and AFS personnel on the bomb scene reported, 'the householders – and the children in particular – thoroughly enjoyed the show':

'It was an awful night, I expect you heard about it; it was in the news. I was up half the night with incendiaries – the ARP couldn't cope, and we all turned out to help . . . we got a packet our way . . . I thought I'd feel like death this morning – but I don't. I feel marvellous – on top of the world.'

'We were coming home (from an ARP training class) last night, and about a quarter of a mile ahead a green flame sprang up, and we knew they were incendiaries! We were as pleased as punch!'

'I found a bomb – an incendiary – outside my house which hadn't gone off, so I rushed indoors and found a shovel to cover the bomb with earth. It was a brand-new shovel; and I was just putting down the earth . . . and it suddenly blew off! It burnt the hair off my hands; and broke my shovel in two! Did I swear at blasted Jerry then! – My new shovel! – I'd been enjoying it until that happened!' (Older clerk, talking eagerly, with much laughter and gesticulation).

'I helped to put out three. When I got in, my old woman pushed a heap of chips in front of me and said, "Eat those!" I never enjoyed anything so much in all my life.'

'I wouldn't mind having an evening like it say once a week. Ordinarily, there's no excitement, nothing to do or anything.'

The last phase touches on a vital point. In those days, before TV, foreign air travel, packaged sunshine holidays and open sexual permissiveness, a great many British found life vaguely dull; more so surely than today, when the same feeling nevertheless remains quite near at least the urban surface?

The blitz introduced high drama. However subjective and confused, a clear pattern emerges, again and again, as far as large sections of the civilian population were concerned, at this stage Londoners especially. The London trajectory was greatly simplified in so far as the metropolis was constantly bombed, so that the people lived for weeks on end under continuous warning – whereas in the provinces, this was seldom to be the case, the experience was much more erratic (and could be bafflingly so), making steady 'learning' of adjustment routines markedly less easy for most people who 'stayed put'.

Without, then, being dogmatic, it seems reasonable to distinguish for London, five phases of *major* adjustment after direct bomb experience, each varyingly applicable under different conditions.

1) (*First minutes*) Some shock: stupefaction: minor impairment of judgement. 'Normal' motivations such as fear, concern for others, anxiety, in abeyance. Physical pain is commonly not felt, and injuries can go unnoticed. Repetitive talking, giving of inappropriate orders, etc. (especially for male family heads, becoming stereotype masculine leaders).
2) (*next 1 or 2 hours*) First-stage recovery; return of sense of reality and 'appropriate' emotions. Concern for others, and for extent of damage, etc. Injuries felt – pain, bleeding, losses begin to be noticed.
3) (*succeeding hours*) Uncontrollable flood of communication, by word, gesture, laughter. Anecdotes, personal experiences; loud claims and counter-claims as to who has had the 'worst' experience. Excited speculations as to extent and nearness of damage. Repetitiveness in both vocabulary and subject-matter is characteristic – a person will repeat his story over and over again, in almost the same words, often to the same listeners. Excitement at this stage is intense, almost at times manic.
4) (*throughout about 48 hours after first-stage recovery*) From the babel of communication, individuals tend to emerge with a sense of intense pride, of enhanced personal worth. This sense of ego-enhancement may last hours, days, or years. A corresponding uprush of pride is also evident on a neighbourhood level – engendering great resentment when radio

or press under-play the incident (this was stronger in smaller provincial centres – see following chapters).

5) (*after about 48+ hours*) Return to 'normal'. The individual reacting with resourcefulness, annoyance etc. to the ensuing material discomforts – absence of water, heat, cooked food, etc. Concern for others and for damage sustained is back to near normal. But a basic 'term' is past, raids seldom hold further equivalent anxiety.

This is an ideal trajectory towards long-term adaptation. No one person followed it exactly like that.

(5) Downs and ups of adjustment

By or before the middle of October 1940 millions of Londoners had 'got over the worst' effects of being intensively bombed. Very few remained in town who had not developed a new normalcy of their own – not a carapace so much as an outer sponge for inner protection. The bomb-baptisms of September had been sufficiently wide-spread for everyone to feel personal identification with escape from death.

Nearly everyone who had stayed above ground in Greater London had shared enough horror, fear and seemingly senseless, random death to sense the realities as something far less severe – but much more irritating or distressing – than the fantasies of the thirties built up by those in the know. Every street now knew the facts of the new war.

The fact that events into winter 1940 were, after all, bearable, gave great relief to the more anxious, reassurance to the rest. It helped, too, that the first course in being blitzed came during an exceptionally fine September. It would have been harder to adapt had it begun in a bitter October. And if, by then, the whole thing was losing its excitement to become boring, that was a state of mind familiar in peace, powerfully accentuated by blackout and the other restrictions of war.

Fortifying the temporary feeling of having entered a brave new world of adventure, with its relationship to the unceasing threat of sudden death, came a kind of pseudo-statistics, hardly more inaccurate than many of the preceding official predictions.

There was, for instance, a widespread belief that once you'd had your first 'near-miss' you were unlikely if not actually *unable* to have another. Associated with this, all through the blitz in London (and then elsewhere) the concept of 'one bombability' – that no two bombs would ever fall in one place – remained unflinching folk-dogma.

Of course, with approaching 100,000 bombs of one kind or another spread over London between 7 September and 13 November 1940 many people had not one but many encounters at some remove. Indeed, the feeling that one was bombed every night, that even remotely distant bombers were bombing London – i.e. you – was another big advantage over dwellers in smaller towns like (say) Southampton, where a raid on the place meant *every* bomb actually much nearer every Sotonian, and next morning's destruction smoking every nose. It was not that most Londoners stopped being worried; rather they ceased to treat individual bombs as necessarily dangerous, even lethal in a personal sense. Another London diarist housewife (husband and young son Kenneth) could still complain at disturbance of the domestic scene on the last night, 13 November, before the blitz switched to the provinces:

> Early last evening the noise was terrible. My husband and Mr P were trying to play chess in the kitchen. I was playing draughts with Kenneth in the cupboard. Mrs P was sitting under the stairs knitting and waiting for baby David to go right off to sleep. Suddenly the raid came closer. The draughts flew in all directions as I covered Kenneth. Presently I heard a stifled voice 'Mummy! I don't know what's become of my glasses.' 'I should think they are tied up in my wool.' My knitting had disappeared and wool seemed to be everywhere! We heard a whistle, a bang which shook the house, and an explosion (but not, we thought, from the bang, or rather boom). Well, we straightened out, decided draughts and chess were no use under the circumstances, and waited for a lull so that we could have a pot of tea.

There were only twenty-five bombers over London that night, one of the lightest raids for over two months; next day came the switch to Coventry ... The important thing was that there was now an established routine for living, however sporadically disturbed – not really disrupted unless by actual physical destruction.

Any steady line of revised habit was naturally subject to its ups and downs, as in any life process. Special nights, sometimes a few moments only, could temporarily shake the pattern, as with the same housewife three nights before (10 November):

Mr Hitler has taken a violent fancy to our area. I don't admire his taste. We passed the noisiest part of Friday night quite merrily! The warden (off duty) came in to ask if we could guess where the bomb had fallen that seemed to land close by and not explode. There was doubt whether it was a time bomb. It was a terrible bash. My head seemed to close down as though the cervical vertebrae were closing like a telescope. It seemed a dead weight on our roof. We heard it coming and had covered the children and were standing over them. The men crouched in the kitchen. We heard later that a lady next door slipped down four stairs.

However many such upsets, however, they rarely changed the post-September family routines. These were commonly so strong that one feels they could have survived anything. Check with the attitude of yet another (West) London diarist. She worked in an office all day, her father and sister working too, the brother unemployed except as a church organist on Sundays. On 10 November like the previous diarist:

At 6 a.m. bombs dropped.

I awoke to hear a roar and thundering, to feel that horrible 'got you' thud of a heavy bomb and the sound of half the world raining down on us. My mother and sister were both giving little screams; I put my arms round my sister and said 'It's all right' several times. Mother got off her bed, crying 'The house has been hit' as she ran to the stairs. 'No it hasn't' I shouted, and then, suddenly felt sorry for her and said 'Poor mother'. 'Get under the bed' said my sister. I didn't want to, because the thought of an iron bedstead falling on me didn't appeal. There was a flash and another came down, and our telephone suddenly started to ring. My sister again told us to get under the bed, so I got under and pulled at her to follow me, but she didn't move. She said afterwards that she *couldn't* move.

Mother screamed for father and brother upstairs, asleep. 'Eventually they both came down, father shouting . . . he wanted to get into bed again . . . he didn't get enough sleep and nothing else was happening,' he yelled. Then she went out and up the road in the dark to see if sister-in-law Mary was O.K. Two houses were gone that side, a third 'seemed to have its roof sliced off'. Mary's home was unhit. *Mud* everywhere – a common winter bomb effect. Suddenly shells started bursting overhead, and (another common effect, this one in personal feeling) – 'we heard Jerry feeling his way back – I suppose looking to see what he had done.'

Back home, the family, rallying once more, started one of those fierce quarrels which were often observed after blitz-crisis, especially as familiarity increased the boredom factor. Our diarist 'beseeches' them to stop. No hope.

Dad turned on me – said I should show respect to my father. He said I should answer for all 'these things', to God one day. I told him that was my affair – I shouldn't have to answer to him anyway so shut up! After that we made some tea – my brother had gone back to bed, but it was time for him to get up; so I took him up a cup and told him so. We all felt a bit better. There was a terrible hole in the bathroom ceiling with sacking hanging through it and mud everywhere.

Came down: the family faced fresh domestic problems of a kind by now familiar all over.

Daylight had overtaken us and I felt a wreck, still having curlers under my turban. Mother was cooking breakfast and everyone had started to shovel away the mud, so I started on our front path. My sister came back and we had breakfast. Then began a hectic day. First we cleaned up outside, swilling down water, dumping debris into the road for the council to remove. Then we began to sweep away plaster inside. There was another big hole on the landing, a smaller one in my sister's bedroom. Then it began to rain, so I put on old clothes and went into the roof. This was very discouraging. There was a huge hole above the bathroom and quite 20 more small ones; one chimney was off and another had a nasty lean. Everywhere were dropping sounds. I put buckets, tubs, bowls under the worst. I rammed a piece of lino over the landing hole and old mats over the bathroom one. Descended to lunch hungry.

When the family reunited at dusk, brother unexpectedly refused to share the usual bed upstairs with father, as a result of the morning's quarrel. So sister made him a place on the front room settee – none of them contemplated changing the stay-at-home habit, though there were public shelter facilities nearby.

We spent the evening trying to clean ourselves. Afterwards I did a little sewing, and mother knitted but my sister just lay on the bed. I had to do something, although I was so tired, because Jerry was dropping more bombs somewhere. Mother said she would sleep with Dad in the dining-room, so my sister and I tossed for her bed; but I lost! My sister bounced up and down on it and said 'The only nice thing that's happened this evening'. 'To you!' I said gloomily and settled down on our usual mattress. We undressed as usual. When I closed my eyes I could see broken roofs, dust falling in on me, and then tiles would fall away and I would give a little jump. The tick of the clock became the drip of water, but soon after 12 I fell asleep from sheer exhaustion.

The whole diary of this week illustrates the distress and minor misery suffered, yet accepted more or less stoically, inside the thankfulness for living on and the necessity for carrying on earning, eating, liking, sleeping. Quite casually this young woman mentioned in her next day (Armistice Day) entry that a fellow worker at her office 'already living in a house *minus doors, windows and ceilings, by himself*' phoned through from a call-box that the front of his house had now been blown in: 'so he wouldn't be in (to work) today'. Regularly, it took that much to keep the blitz-trained from their daily jobs for the whole day.

The following days for this West London family continued domestically disrupted. They tried to get a local council official to inspect and assess the damage. No use. Eventually they found a private contractor, who appeared in heavy rain at 3.30 p.m., got soaked up on the holed roof, went home fed up, without fixing (or saying) anything. They tried again themselves with an old tarpaulin dredged up from somewhere, trying to keep the house tolerably dry. After a long saga of struggle to get council help, they found a local decorator to show them 'how to tap holes in slates' and secure a roof patch. Presently he was himself moved to help:

I changed into my best frock – tried to do my hair and powder my face and mend stockings – my hands were rough and hard and stiff and just wouldn't do anything – everything fell under chairs or couldn't be found. We were all in a state. So we got tea, *toasted crumpets.* Our decorator came in for a cup, and he and mother started to discuss which of them had been in the district longer. He has been here 31 years – mother 32! We parted the best of friends – we know for a fact that he skipped one of these upstarts who've only been here about 15 years to come to us!

They were lucky, as a belated official visit demonstrated next day, at last . . .

The council men called as they promised, and told mother they hadn't any slates. She asked them to go up and first aid it – they are *the first aid squad anyway*. They went up, swore at it, and then went away without touching it, or knocking at the door again. When last seen they were mending one of the easy roofs with two or three holes over the road. We have heard of two cases where they have actually removed the lino that owners had laboriously put over their holes, and taken away loose tiles, making the holes bigger.

That last London night, 13 November 1940, trouble struck afresh inside the home:

Had a meal of baked beans on toast, and afterwards we were feeling worn out, but had to start clearing away to make the beds. I made my brother help me with Dad's bed, while Mum and my sister made ours. Then we set Dad and my brother to wash up. Mum went into the front-room to make my brother's bed on the settee, and gave out a cry of horror. There was water dripping from the ceiling. She dashed upstairs and started rolling up the soaked carpets, my sister followed with cloths and basins, I bounded into the roof, seized the tarpaulin lent us by our neighbour, and hauled it through the trap door into the stricken bedroom. We spread it right under the patch, where the paper had stripped off and water was running out of a long crack in the plaster. I tied the strings on the edges to a bed-end, and window knobs, washstand, etc., so as to create a puddle; and we set dishes and a bread bin under the actual crack. Then we found there were two holes for tent-poles in the tarpaulin and a small slit, so we had to set bowls underneath the tarpaulin to push it up at those spots. There was a merry sound of dripping into the contrivance when we'd finished. Then we started to readjust the furniture – put them in dry spots, covered them over etc., and spread overcoats and an oilskin cape underneath another suspicious-looking patch.

It was well after midnight before we could drag ourselves into our beds. Slept the sleep of exhaustion.

But they managed; within the week they had things well in hand once more. Meanwhile, far across in the south-east corner of London (S.E.18) another home, Anderson-shelter-orientated, was holding together with difficulty largely because of internal sexual tensions. Their Anderson in the small back garden was typically damp, but never flooded (many were). For reasons not clear, the husband insisted they both used this shelter for *all* alerts, day and night. Yet he always brought his wife back indoors at 5 a.m. – 'no matter what is going on'. This maddened her. On 6 November they had a big row, mostly because she would no longer shelter by day (realistically enough) and only slowly, reluctantly, by night. On two following nights they fought. By now she was utterly bored with the tiny Anderson, plus their dog. Feeling the relief of the switch to Coventry on 14 November, when (159) bombers returned to London on 17–18 November things came to a head:

Evening raid began so early, 5.30. Caught me unready, I thoughtlessly asked could I do a bit in kitchen before going to shelter. He kindly helped with bottles and we went down. I was terribly fed up, so much so it was all I could do to keep tears back. Nervy really and just could not bear the thoughts of having to stay cooped down there for 14 hours. Dog had not been walked either but I dared not say about that. Of course, we ultimately rowed. H, nervy himself, finding my gloom and taciturnity too much for his irascibility. It was a ghastly row too awful to retell. I foolishly said I did not care if I did get bombed. H, perfectly furious, replied, 'What about him then', or words to that effect. After that I was told I could do what I liked, he'd stick it till after the war but from now onwards I'd get no affection or consideration from him. He knows how to get to me every time. I begged pardon most humbly and H stormed out into the barrage. Having wept my heart out I followed him. As I write, can't remember how the night was finished. I so distraught and unhappy. More so than I can ever remember for H also said I was losing all sense of love making and *intuition* about it too.

They made it up. By 8 December (413 bombers) she had accepted the Anderson as her fate:

Had our meal, I washed up and about to do night jobs when raid got very fierce. We did a bunk to shelter, dog not walked either. Thought to go up later and take him out but no opportunity.

Terrific noise. We tired and went to sleep. I woke about 10.30 p.m. and heard 6 bombs and then a D A.* Pretty close by sound. Went to

* Delayed Action bomb, in practice not distinguishable from any other unexploded bomb.

sleep again. Gunfire and noise terrific after midnight. Only dozed. Then after one, swish, swish, all over place. We crawled out hastily. None near us but all sky behind us brilliant from many incendiaries.

Hubby went over the fence with a spade, incendiary hunting. Fires all round. Settling back in the Anderson, when dawn finally brought them out: 'Found it overcast and showery ... Both a bit weary.'

Such were the milliard minor crises – and triumphs – of living through that blitz. The capacity to adjust and endure will only surprise those unaware of the human will deployed in Dachau 1945, the Sahel today, among the Arctic Esquimaux and the nomadic Punans of equatorial Borneo. As that West London housewife put it, movingly, in mid-November, indicating incidentally the single fact that in no family could everyone get fully used to blitzing and stay that way every night, even after weeks:

At about 10 o'clock things began to hum, and developed into quite the worst raid we've had yet, with a terribly noisy barrage. We didn't attempt to go to bed till midnight; we were all very much on edge. My sister and I got into our bed and my brother into his room – and never came out again that night. But my father was really unashamedly jittery *for the first time*, and neither he nor my mother made *any attempt* to go to bed. Once or twice I almost dozed, but some suspicious crash, or vibration of the floor-boards would wake me every time. Finally there was a series of horrible whistles overhead, which put the lid on things. There were no explosions. My father started to look for incendiaries (it turned out he was right) and went to the door about 6 times in his pyjamas. Then things were a bit quieter and twice he went down to his bed in the dining-room; but each time the barrage immediately burst out again (sometimes it seemed to be drumming out tunes, like an enormous Boys Brigade, drum and fife band) and he came trotting back. Finally mother got him to share her little single bed. She herself wore all her clothes. At the first attempt to arrange themselves, she moved over too far and fell off the bed on to me! Finally they managed somehow, altho' Mother swore he had all the bed and slept much more than she did. None of us slept till about 3 a.m. and even then I could still hear the guns thudding away through my sleep.

The milkman, as usual, that morning, called out: 'A wicked night.'

(6) *'Non-adaptation' in town*

The disciplines of London living, all self-imposed (in glaring contrast to life in the armed services), were rugged upon the ego, difficult to maintain through thick and thin. The breaking point, invisibly, was seldom far round the corner. But then, no one quite knew *how* to break out; how, for instance, to show 'defeatism', should one actually feel defeated. Not that this is a purely wartime, let alone purely blitz pressure. The same ache works inside many seemingly normal people in everyday life: the urge to chuck the lot and disappear. Only there is no reality mechanism, no bridge from fantasy to achievement of a kind equivalent to the release provided by the 1940–41 blitz itself. In war, too, the sanctions on staying steady are stronger in terms of respectability, good citizenship, the due observance of familiar and wider obligations.

To retreat into a species of personal, private defeatism was the other way out, taken more readily if you could rely on someone to see you through the practical part of life. Some old people opted half-out like this. Some young and some old, some females and fewer males, never could get over a near-allergic reaction to the whole bloody business. But they were astonishingly few in number: astonishing, that is, in the face of the pre-war expectations. Virtually 'none' of them carried the inner retreat so far as to go – or report – crazy, to drop out of the home altogether and head for any one of the many thousand beds hopefully prepared for the expected flood of such mental casualties.

Although our evidence suggests that those with any degree of *lasting* incapacity to adjust were very few among those living through the London blitz, the plight of these, if they for some reason could not leave, was pitiable indeed.

'I can't bear it, I can't *bear* it! If them sirens go again tonight, I shall die!' sobbed a 40-year-old Kilburn woman on 17 September. This woman (whose home had suffered no damage this particular night) was showing up far worse than she had after the first bombs in this area, when several of her windows had been broken and a close neighbour bombed-out.

'It's me nerves, they're all used up, there's nothing left of me strength like I had at the start,' – was the way another described the sensation of failing courage as the bombardment continued.

'It's the dread, I can't tell you the dread, every night it's worse,' – said a sixty-year-old working-class grandmother. A middle-aged construction worker described his wife's plight thus: 'It's getting more than flesh and blood can stand – it just can't be endured, night after night like this. My wife, I've got to get her out of it, she's getting like a mad woman. As soon as the siren goes, she's like a mad woman.' An observer described the case of another, younger woman as follows: 'During the day she keeps saying "I can't bear the night, I can't bear the night. Anything like this shouldn't be allowed" . . . As it grew dark, her state became worse. Eventually she was trembling so much she could hardly talk. She ran upstairs to the lavatory three times in half an hour. Finally when the warning came, she urinated on the spot and burst into tears . . .' Sometimes, the distress was compounded by the fact that the sufferer had been exceptionally unperturbed at the beginning, and had perhaps shown-off about it.

This unpleasant syndrome – of progressively worsening terror with each new experience of danger – could affect A R P personnel at their duties, just as much as the general public. A Hampstead observer described such a case among stretcher-bearers at the depot:

> . . . there was a complete change in D now (10 October). In the early days of the blitz he used to enjoy going for walks in the moonlight in the barrage, watching searchlights etc., and often went out on to the balcony of his room to watch raids at night. In the early period, he was one *of the least nervous people* I met. Now on his off-duty days he goes down to a shelter in a friend's garden every night at dusk, says he can't sleep at all in a house. A month ago he used to sleep right through the most intensive raids and barrage.

As has been said, only a minority were affected in this way – though it must be remembered that by now something like a quarter of London's population was gone – many of them, no doubt, self-selected as 'not being able to stand it'. Had all the six million Londoners been compelled, like conscripted soldiers, to remain in the firing-line regardless of individual idosyncrasies, then the minority just described might have been much bigger.

The majority who stayed found – mostly to their own mildly delighted surprise – that they were much tougher than they had earlier thought. That added strength for staying. A middle-class woman in her mid-thirties, mother of two, fairly reflected a frequent personal experience.

Suffered agonies of fright at the time of Munich, and again at beginning of present war. Lay awake nights with visions of appalling devastations. Before war started, fled with children to Scotland to escape from the horrors expected immediately: would never have returned to London if the blitz hadn't seemed to be indefinitely 'off'. Sweated with fear at the mere idea of being woken in the night by sirens, let alone at that of bombs and shells; was constantly imagining car and other noises were sirens or guns. Convinced of being most utter and complete coward – very ashamed. Sure I was bound to panic and lose control if things got hot . . . Now find myself almost completely proof against fear and jumpiness . . . am the most fearless among my own circle of friends and acquaintances . . .

(7) Death

Sam had the nearest escape from death he will ever have. Yesterday a friend of his told him he was going into the army so Sam said 'Meet me tonight and we'll have a drink on it.' It was the first time he had ever ventured to go out at night – he set out in a quiet spell – a beautiful moonlight night. Two bombs suddenly whistled down and fell about 100 yards in front of him; he saw a sheet of flame and then mud and debris were sucked up into the air in a whirlwind. He flung himself flat behind a low garden wall and the next moment a bomb fell just round the corner. Said Sam, 'A man in a tin hat came and asked me if I was all right and I stammered yes because I didn't want to be carried off by them, but my knees were shaking so much I couldn't stand up, and for about 20 minutes I couldn't speak properly. When I had regained my faculties I went straight home again, and never had my drink!' He is not going out at night again!

Sam, in that London diary, was lucky. Beyond all questions of adaptation lay the shadows of those 13,596 Londoners killed by enemy action before the end of 1940, plus another 18,378 consigned to hospital. But this is a study of the living. The dead keep no diaries.

Quantitatively, death composed a small part of the same picture, a corner. The services for dead and injured had been well prepared, too. Since the casualties were far less than expected, this part of the post-blitz machinery did not come under great strain like the rest – like, for instance, those services for the more than a million households with homes to some extent damaged by bomb, the one in five rendered at least for a short time homeless.

It is proper to notice, with respect, the dead, who might otherwise tend to be overlooked in this account of the living. As with those who die in the rhythm of ordinary times, they tended indeed to be looked past, to be put aside from continuing concern by those not directly concerned.

Under the potentially fearful conditions of being blitzed, the extent or frequency of references to death seem remarkably small. The record shows no special discussion of the theme, no new metaphor or concern for the novelty of sudden demise from the skies. One observer, working very close to the corpses in a hospital, wrote after a September raid:

> Didn't know where to start, got my tin hat and gas mask to look O.K. and show I knew something of what was wanted. Somehow got on to stuffing a corpse (old woman) on a door, very heavy with three other people taking it out of first aid post round to back of hospital towards the garage where nurse (or sister) had said, on the way met 'Eddie' young porter who said that he was 'in charge' of the mortuary. He said fill up the mortuary before garage, so slowly onwards to mortuary. I very tired suggested a rest after gone a way (uphill) – gratefully accepted, one of the men saying we should dump her anywhere, as she was beyond help, go and help the wounded. Nobody replied tho' I agreed with him mentally but thought it better to go up to mortuary – not to leave old girl in the open. Then Eddie thought of barrow, got it, and we put her on it: Eddie and I took her up to the mortuary: fetched out a metal 'marble slab' on wheels, rolled her off the door on to it, she being covered by a thick red curtain bloodstained: took her into mortuary where already 4–5 corpses in similar bloody condition. Took back barrow to F A Post, dumped it near door and looked for more work.

And so on, for pages, rather 'heartlessly'. One is reminded of Tolstoy's sentence (in *The Death of Ivan Ilych*): 'The awful, terrible act of his dying was, he could see, reduced by those about him to the level of a casual, unpleasant, and almost indecorous incident (as if someone entered a drawing-room diffusing an unpleasant odour).' To change the metaphor somewhat, the normal human capacity to sweep death smartly under the carpet was if anything accentuated by blitzing: it was no good wallowing in it, though naturally few would not respond to a good funeral. This effect, probably desirable in terms of psychological health, was made easier by the aforesaid efficiency of the services in disposing of the remains.

These services eventually suffered their own direct hits. A medical student sleeping in at St Thomas Hospital was doubly involved in a typical incident of this kind, when the place was hit near the start, on 8 September. He had a bed in a basement corridor that ran the whole length of the hospital, was awakened in the early hours by 'an appalling series of explosions'. Gassiot House, a Nurses' Home, was 'blown to bits'.

With another student, he searched the debris in a deep cloud of darkened dust, went on to the adjacent women's ward which was being evacuated in the dark – the windows all blown out. He carried patients to another ward. They were 'marvellous'. Some joked, some stayed silent; a single case of hysteria. A new bomb landed nearby: 'stifled cries of fear and dismay'. Tact, good humour and 'a large quantity of morphine' maintained tranquillity.

'After this,' the student wrote, 'there was nothing much to do.' So he went and 'looked again at the wreckage', then to the canteen for two glasses of lemon-barley to raise his blood-sugar. At dawn he went home, bathed off the debris and slept in an armchair until 9 a.m., then back to the hospital by bus. His diary summed up the whole exercise in a student perspective relevant at the time:

This was the finest piece of 'muddling through' it has ever been my lot to witness. No plan or organization existed in the hospital to cope with an emergency such as this; normal routine was simply thrown to the winds. But in spite of this there were no hitches in the proceedings, for the simple reason that a suitable person was always present with suitable equipment at the right place at the right time. This was due entirely to sheer guts and initiative and intelligence on the part of nurses, staff and, especially, students.

At no time did I see evidence of panic, either among staff or patients. But nervous tension ran pretty high some time after the bomb had dropped. While I was sitting in the canteen the gas ring 'popped' loudly; every person in the room started nervously, and, seeing what was the cause of their fright, began to laugh.

At the other, private end of the life–death cycle, it is not possible to detect, anywhere in this record for London, any sustained turning towards God or faith in – or hope for – an after-life. On the contrary, plenty of people seem almost consciously to have turned the other way, as if the whole thing was against the normal decency of death, faith and divine purpose. A London woman put it another way in a dream she reported for January 1941 in which she had been given, in the bath, an Automobile Association map, along with a notice which read:

In view of the recent blitz, it would be useful for all members of the A A to be familiar with the roads leading out of London and their ramifications.

Especially is it necessary in view of the great importance this Association holds of the doctrine of the Resurrection of the Body, in the light of recent air raid deaths. This doctrine extends to all domestic animals including cattle (we are not sure of sheep and goats).

(8) Noise and sleep (the key)

One pre-war experiment with anti-aircraft fire on a low-flying target gave only two hits with 2935 rounds fired. A A command as a whole, by autumn of 1940, claimed 2444 rounds per night-hit with heavy guns, and a September figure of 1798 shells for each enemy aircraft destroyed by A A weapons of all calibres. But, as the official historian of this service quietly remarks, these claims were 'nearly always exaggerated, sometimes grossly'. Interestingly, this seems to be a phenomenon common to *all* air war claims, on both sides.

Millions of Londoners neither knew nor cared about such statistics. Here, as elsewhere, throughout Britain, nothing distressed people more – in the first phases – than the feeling that they were easy targets: the idea that they could be bombed deliberately, selectively, was emotionally far less acceptable than the idea that the other side could only pour it on indiscriminately, with equal mischances for all. Nothing enraged more than a seen (or felt) failure to make things as difficult as possible for the bombers. To sit or lie in one-sided passivity was too much: this was probably a big factor in the panic caused by relatively mild air attacks earlier in the Middle East and elsewhere, from which general conclusions had been so confidently drawn by the Douhetists.

Government, through AA and RAF, responded to this immediate, loudly-expressed requirement by loosing off a liberal barrage of night fire, and as much as possible night-fighter support. By mid-September 1940 a survey of Londoners could conclude that the increased barrage gave: 'the greatest pleasure ... highly emotional in content, and almost unanimous'.[4]

Men were the most enthusiastic, old and young.

'I must say that though the noise was awful last night, we were all relieved to hear it. The louder it was, the greater confidence we had.'

'You can't sleep with the guns, but it's a good sound.'

'I love the sound of the guns!'

A young man, who claimed that he found the sound 'extremely satisfying', went into loving detail:

There are three distinct sounds from AA fire. First, a crisp, clean explosion, followed in a fraction of a second by another much smaller similar sound. In our household we have assumed this to be the noise of a fairly near AA gun with its barrel facing the house. Second, a more muffled explosion, followed by less crisp rumblings. This we assume to be the sound of the same gun pointing away from the house. The rumble is assumed to be echo. These are followed two or three seconds later by the same metallic sounding explosions, assumed to be the bursting of shrapnel.

Women were less enthusiastic. Some were positively anti – with remarks like: 'It gives you a headache, all those guns.' Or 'bloody row, it's worse than the bloody Jerries' (a 60-year-old trying to get her grandson to sleep in their Anderson shelter). And 'them damn guns, I could kill 'em' (a charwoman plodding to work). Annoyance was mixed with satisfaction, noise normally the lesser of the two evils. As ever, there were wide extremes.

One young woman, told by a neighbour that the guns 'had

probably saved Battersea Power Station' exploded: 'Battersea-Bloody-Power-Station, if that's what they're trying to get, why don't they let 'em bloody *get* it and be done with it? Then we might get a bit of peace here!'

Then there was the 35-year-old chauffeur who (as reported by a workmate) hated the gun noise so much that he started cementing the big fireplace downstairs – 'there 'e sat, going 'ammer, 'ammer, 'ammer, you couldn't hear yourself think.' For nearly everyone, hearing was the most important sense used. We shall never know whether those with acute hearing suffered less or more; probably it was far more complicated than that. But in the final count, most considered the noise of the guns to be worth it, while those who couldn't stand it could go underground or use ear-plugs. Most soon got used to the uproar. Usually within a week they developed something near a protective deafness on the threshold of sleep. Personal anecdotes could soon be heard, at random, all over town:

'You know what, I woke up this morning to find glass all over my bed, and all my books on the floor, and I hadn't heard a thing!' (Young woman).

'At the start, I couldn't sleep at all, waiting for the next one: but now, well, the most I do, I might dream that the windows have been blown out, and when I wake up next morning, they *have*!' (Young woman).

'The wife and I, we were blown clean out of our beds on to the floor. We slept through most of it, and next thing we knew, we found ourselves the other side of the room on our hands and knees.' (Grocer's delivery man).

Sleep, uproar regardless, became the key to all blitz acclimatization.

People who had previously fussed about road sounds or snoring found they could learn to sleep through 'anything' and come out rested. With sleep, nearly everything else became at worst tolerable and at best fun. Once those Lyons Corner House waitresses began to sleep better again, there was no need to be so late for work.

Early on in the London Blitz it became clear that loss of sleep was the major problem, outweighing all other distresses and anxieties:

'It's not the bombs I'm scared of any more, it's the weariness . . . trying to work and concentrate with your eyes sticking out of your head like hat-pins, after being up all night. I'd die in my sleep, happily, if only I *could* sleep.' (Civil servant, female).

'We can't go on like this, we *can't*. Last night, I thought, I can't stand this, I'll chance it, you can only die once: so I come in again (from Anderson shelter) and went upstairs to bed. But it was no good, I couldn't even doze off, all that noise, and wondering about the kiddies out there, so I came down again. How can we go on, I'm wondering? How *can* we?' (Mother of three children, who slept soundly all night in Anderson).

'Sleep? I don't think I could sleep now, even if the bombing stopped tomorrow. I'm all – you know – I haven't slept for five nights. It's gone from me, I wouldn't know how to lay me head in peace any more . . .' (Labourer's wife).

Sleep – or the lack of it – almost replaced the weather as a topic of conversation; strangers in the street would greet each other:

'Tired this morning?'

'Dreadful, isn't it?'

There were, of course, many other, lesser adjustments to be made once nocturnal bombardment struck, even if one could virtually ignore the noise of night and mess of morning. A few bombs could deplete or disturb the amenities, already inevitably curtailed. A continuing survey through September centred on six kinds of London shop and six leisure outlets, plus pubs, restaurants and cafés, churches and libraries, as well as in the home.

Most places of leisure were 'badly hit' by human abstention, accentuated by gas and electricity difficulties. Many shut down. The amount of beer drunk showed no noticeable per capita decline – only it was more heavily quaffed before darkness. Church activities, mostly diurnal by now anyway, showed no marked change in London. Most Anglican places of worship closed directly a siren sounded, however. Some were locked all day. Reading had declined, as one of several losses from sheltering or huddling in an ill-lit space. Radio listening decreased likewise, though about a tenth of listeners – as also of newspaper readers – increased their activity as an antidote. Many had eating and mealtime habits altered, though over half carried these on more or less as usual. The final overall conclusion: 'Most noticeable is the way in which people have tended to give up activities and *not replace* them. *Doing nothing* has increased (alarmingly?) as a leisure pastime in days and nights when people need, more than ever before, something to do, something creative or relieving or soothing.' Government and local leadership took no visible interest in these aspects of maintaining the public mood. Some of the apathy was a form of protective withdrawal; much more due to tiredness, a desire to rest when sleep itself was difficult.

Many people also kept records of their personal sleep patterns at this time, while we asked others, the less literate, in simple samples. First, what did they (try to) sleep on?

The men of the home, still very predominantly wage-earners, got some preferential treatment that September. But a high bed rating was due mainly to a masculine minority staying determinedly upstairs while wife/mother went down and camped with the children.

SLEEPING PLACES, *London, mid-September 1940*
(as percentage of each sex)

	Male	*Female*	*Both*
Bed	42	19	31
Mattress (on floor) couch, or cushions	39	46	43
Deckchair or other chair	14	23	18
Miscellaneous	5	12	8

The stayabed habit was already sharply on the decline by 12 September, and reached a low of 17% (both sexes) by 16 September. The very marked tendency here was a recognition of the continuity of the blitz and the desirability of setting up a regular family menage, preferably downstairs, as a means of preserving home life. So, by late September, although the number sleeping upstairs had declined, so had those using their own or public shelters – from a high of 44% at first, to half that and lower. Many shelterers moved indoors just as upstairs moved down, to make over half the town in familiar units camped about the hall, kitchen or front room (mostly the hall and 'under the stairs').

It was certainly easier to *sleep* downstairs than outside. This was the determining factor for most people, outweighing other considerations (not excluding fear) as tiredness built up. The extent of this tiredness was great.[5]

SLEEP ACHIEVED LAST NIGHT, *London, 12 September 1940*

Informant's sleep statement	*Percentage within category*
'None'	31
Less than c. 4 hours	32
c. 4–6 hours	22
More than 6 hours	15

The last figure had already risen to 22% by 16 September, as the downstairs movement grew. Post-war research, notably by G. S. Tune of the Medical Research Council's Unit at the University of Liverpool (1968), working with a sample of 240 'normal adults', has shown that the commonly recommended eight hours is not normally necessary. Women tend to sleep less than men, wake more erratically – as so markedly in air-raids. Daylight napping readily compensates for night loss, especially among older people. Studies by J. P. Masterton (1965) also indicate that stress may enforce shortened sleep even among children, to such an extent that a 'sleep debt' is accumulated, carried forward over quite long periods until 'paid off' by holiday or lull periods. London was inhabited, that autumn/winter, by sleep debtors, with relaxing holidays hard to find. Rather, people adjusted by gradually immunizing themselves, as shown in our changing figures for those claiming *no* sleep (as above):

	Date (1940)	*% saying they got no 'real' sleep*
September	12	31
	15	28
	19	24
	22	9
	26	7
October	6	3
	16	5*
	30	0
November	10	0

***(15 October was much the heaviest raid of this London series, 410 bombers – compared with average 163 nightly; see next table.)**

Where people got less than four hours over several consecutive nights, it began to show. Adjustment continued to improve steadily, however – and there are always chronic insomniacs, including some who chronically say they can't sleep when they can.

SLEEP (5 BOROUGHS), *September–November 1940*

	Date	*% sleeping more than 4 hours*	*Scale of attack over London, previous 3 nights*	
			Planes over	*Tons H E**
September	12	37	370	450
	19	23	740	875
	26	57	705	815
October	6	77	355	490
	16	83	865	1090
	30	93	445	540
November	10	94	445	500

Mothers took lack of sleep hardest, despite a better capacity to wake, sleep and reawake. Their first concern was nearly always for the children, then husband (wage-earner), then any older folk. This is how it worked out for one reasonably typical family, parents with daughter (age 6) in Fulham:

In peacetime the child went to bed at 8, the father at 11, and the mother last of all. They rose in reverse order between 7 and 8. Since the raids they have all slept in their Anderson shelter, the mother and child on a mattress, the father on a deckchair with cushions and blankets. The child goes to bed an hour before the warning is expected, and sleeps for three hours after the 'all clear'. The father, his wife alleged, would sleep through an earthquake, and goes to bed at his normal time. She herself does not get much to sleep at night but has an hour or two in the afternoon while the child is out playing. On the night previous, the mother had spent most of the night trying to keep the child asleep, and had been fairly successful. Even during the worst bangs the child had not completely woken up. The father had wanted to take over her job, but she told him she could sleep while he was at work whereas he would grumble because his work wasn't done properly. As a result, the father had had four or five hours of genuine sleep, and the child does not suffer at all from loss of sleep. She does what she is told and sleeps when she is tired. The mother is the worst affected but makes the best of it. 'I'm lucky, I manage to get an hour or two in the afternoon; not everybody can do that.'

* (Figures recalculated from basis of Collier: 494–5.)

At least one London diarist snatched minutes of his sleep off and on all day – in his case, in his surgery, between patients. These stratagems went on all over: 'I write up something on the blackboard for them (his pupils) to copy out, and then I let myself nod off until they have finished ... It's funny, I always manage to wake up exactly on the dot ... they never notice.' (ARP instructor). 'I sleep on the train. On the buses. Even waiting at the bus-stop sometimes, or in the lift going up to my floor.' (Female worker).

A middle-aged man claimed to have acquired the knack of falling asleep whenever anyone spoke to him and waking up just in time to answer them: 'Just like any other conversation – you'd never know the difference!' And a resourceful young ward-maid dating her boyfriend every night, by day in a large South London hospital (starting 6 a.m.): 'What do I do? I catch a few winks when I go down for the coal! I have to bring up 22 scuttles, see, and each time I go down the cellar I stand there a couple of minutes, leaning up against the wall like, until I feel myself going ... It's surprising how it sets you up, just a couple of minutes like that, each time you go down like ...'

(9) A question of ear-plugs

The conflict between sound and rest was for a while formidable.

'I want to hear.'

'The noise doesn't trouble me any more.'

'I don't like the feeling of being bunged up.'

These were reactions to an official effort at bridging the sleep gap. 'Not so much to obviate physical damage to the ear as to prevent the shattering effect of noise on nerves' went the characteristically inappropriate official explanation for the distribution of ear-plugs, with leaflets, on 27 September 1940, piloting a sponsored scheme for a form of the Matlock-Armstrong device on tapered rubber with a rim one end to which a string fastened so that the gadget didn't get stuck inside.

By then, about one in ten of London householders had already bought or tried plugs on their own account. But, as we wrote (too simply) at the time in a memo to the future Lord (Dr Solly)

Zuckermann (then at Oxford), even among the early buyers there was 'a widespread non-use of ear-plugs, mainly because people want to listen to possible death'. We went on to suggest that it might be 'stupid to spend a lot of money and effort in distributing anything free and official without at the same time making people feel interested and thankful for the action'. The plugs were in fact handed out without adequate explanation or leadership. But they certainly showed official concern, if only for shattered nerves.[6]

Public response was apathetic. In some places an insignificant number responded. No more than 6% used the plugs after a first experiment. Much of the response was decidedly negative: 'Oh, I couldn't *bear* them! Nothing would induce me even to try them – it'd be absolute torture to me not be be able to hear what is happening. I should go quite mad with fear. I can't bear even to have the wireless on during a raid. I wouldn't put in a pair of those things during a raid for all you could give me!' A woman ambulance-driver (a proselytizing convert to ear-plugs herself) thought otherwise: 'It makes me furious, these neurotic women, they do nothing but complain about their sleeplessness and nerves, and then won't try the thing out . . .Too silly!' The common view of both sexes was overwhelmingly against any hearing un-aid: 'I had a pair at the beginning . . . but I feel I want to know if anything happens – there are the children to see to and all! Take away your hearing, you're lost.' Listening could be very important to one's peace of mind. An elderly worker put this with bewildering clarity: 'Give me the use of me own ears, any day. With your own ears, you can hear if it's quiet, and you know it *is* quiet, and you can rest, like. With them things in, you'd never know if you was copping it or if you wasn't.'

There was, equally, a wish to hear the sound of bombs falling. Although this tended to diminish with time it remained strong:

'It's better to hear what's going on. You can't hear the whizz (of a bomb coming near) if you've got them in.'

'I like to know whether they're near or far from me, and whether they're coming or going – then I know whether I ought to be lying on me face under the kitchen table, or whether I can take me comfort on a mattress.'

Said an ARP worker, speaking of her colleagues: 'They absolutely refuse even to try them; they say they'd go mad, feeling anything might be happening unbeknown to them.' Perhaps they would have gone mad? No research had been done on the psychological effects of blotting out the sound of danger while leaving the danger ubiquitous. We can at least say with certainty that over 90% of the population preferred not to have their hearing subjected to the experiment, although our evidence indicates that the plugged élite averaged nearly two hours more sleep per night of use.

As Angus Calder has properly noted: 'except in the depths of tube, no one escaped the noises of the blitz.' Richard Titmuss, writing in another wartime connection, has a remark much to the point: 'threatened with death, moral aloneness becomes to a man even more intolerable than in peacetime and perhaps more hurtful than physical isolation.' The Tube and similar shelter systems substituted human voices for the guns and planes. That did not (as Titmuss was inclined to think) mean ignoring danger and death. It meant, instead, acknowledging the inevitable possibility but evading what, to some, seemed immediate probability. But few Britons stayed up top in town if they fully believed a German bomb was likely to hit them.

You did not need to lose all fear under these circumstances. You had to learn to live with a new fear added to the older, everlasting ones of civilized man. If unable to adjust thus, physical escape downwards remained an alternative solution to abandoning the whole scene laterally, by running away.

(10) The bowels of the earth[7]

London's elaborate Underground rail network had been rushed by up to 300,000 'panicky' East Enders according to the reports in the First World War. It had to be taken over again, more gently but by force – the force of social pressure – before the authorities would recognize a need for deep shelters in the Second World War.

No other civil act of this war has earned more frequent (and superficial) comment in those histories or documentaries which deal with the Home Front. For it was a conspicuous operation performed nightly in public on the platforms and by day with queues outside the stations, waiting to get in and down, all day, any weather. Any war correspondent would see this, while the vastly more sheltering in their gardens or kitchens went unnoticed.

On 24 October an observer joins a typical queue to share the by now daily routine of thousands. A damp, windy day, there are already a hundred queuers by 10.30 a.m., some of whom have been there since daylight. One-fifth are children; a few men, mostly shabby and elderly. Beside a nearby wall stand rows of perambulators and pushcarts, some piled with bedding, in one a torn grey rug, a faded green and pink patched eiderdown, some stained striped pillows tied with a thick string, a pair of strap shoes and a man's old overcoat. People stand or sit on their bundles, quietly. Many women wear hair curlers.

Morning winds on into afternoon, the queue growing. At 3.30 p.m. comes a sudden liveliness. Several run forward, while others shout at them angrily. The station doors are opening! A dozen in front burst through 'in a sort of headlong rush'. The rest file slowly, police supervising (mildly). Inside, most go at once to buy the tickets necessary for platform permission, though others enter without – they will pay later. 'What time did *you* get here, you sod,' yells a woman lugging a huge bedroll and a child, to a pale young man hurrying in ahead of her.

By 4 p.m., all the platform and passage space is staked out, chiefly with blankets folded in long strips lain against the wall – for the trains are still running and main platforms in full use. On average, one woman or child guards places for six thus marked. A grandchild, aged 8, who has been on queue since 6.30 a.m., explains she hasn't had her dinner yet, but likes the place: 'It's all right on the platform, it's warm. It's cold by the stairs.' The long wait for a draughtless place is worth the effort.

Further down the platform a quarrel is going on. A little boy of about nine is standing tearful but defiant: four or five women are talking indignantly on his behalf. He had got down first and booked places for his whole family. Then a porter had come and pushed his things away to give the place to an elderly woman and her daughter.

'It's wicked, do a thing like that! Freddie started waiting *half past five this morning*! He was up there, with the guns going and all, before any of us.'

'He has that place every night. They've no right to do a thing like that . . .'

'You wait till that man comes back (the porter). I'll tell him . . .!'

The two interlopers are standing by, rather embarrassed.

'Well, I didn't know . . . He just gave me the place,' says the younger woman.

'No, it's not your fault.'

'Oh, no, it's not her fault. It's *his* fault. He oughtn't to do a thing like that.'

'It's wicked. We stand out there in the perishing cold all day. It's not like we did it for ourselves, it's for our husbands.'

The porter reappears. The women all begin talking to him at once. He shouts them all down: 'That place wasn't taken! If I give that place to anyone then that's their place! I'm not going to hear any more about it!' He stamps away and disappears through a door, slamming it behind him.

'Never mind, dearie,' says a woman to the little boy, 'they shan't take your place tomorrow. I'll keep it for you if I have to take my skirt off to put down.'

On to these contested platforms and cool corridors the rest of the families come crowding with darkness. The highest estimate of total Tube population for any one night was 177,000 (or more than the population of Southampton) on 27 September. That would be less than 5% of those left in London. In early November a census of all London's sheltering styles gave roughly:

4% in Tube and equivalent big shelters

9% in public (mostly street-type) shelters

27% in own domestic (mostly Anderson) shelters

Allowing for duty, work and special factors, this left over half staying – as they nearly always had after early September – indoors in their homes, downstairs or up.

The authorities had feared that the Cockneys, once admitted underground, would become a race of troglodytes, useless to the war effort, a menace to health and a source of endless potential trouble. In fact, people left when told and obeyed the porters without the intervention of police. The winter of 1940 showed a low rate for 'flu and other illnesses which could easily have been transferred in the, at first, near zero hygiene of the Tube (for naturally no preparations had been made in that direction). When a few large stations, like Liverpool Street, were left open to all-day stayers, few took the opportunity – less than 150 families in one of our observations.

The undergrounders were not a pack of cowards. They were all sorts, with many reasons, not least the utter inadequacy of protection in the then 'slum' areas. A considerable minority had feelings, maybe obsessions, about their personal involvement in the blitz, like this man: 'I imagine a bomb dropping *straight* on top of me exploding. I know it couldn't really be like that, but that's how I feel.'

Further along this line, near paranoia perhaps, the housewife who feels the Germans are after her personally: 'I like to go shopping *when it's busy*, so they won't notice me along the street.'

While another lady assured an acquaintance that she had given up hanging her laundry on the line because 'it's too conspicuous, all that white!' Such as these could find blessed relief in the bowels of earth.

Undergrounders were not necessarily extremists. They did not all fear less or sleep better than in their own kitchens. Systematic observations at several stations in late September suggested an average night's sleep underground as being 4½ hours for men, 3½ hours for women – largely due to frequent waking breaks, though these diminished markedly after 2 a.m. What mattered more was the general feel of security, in a setting far removed from 'moral loneliness'; a setting as unlike anything previously experienced in domestic routine as the blitz itself overhead. The sociable mix-up appealed to some. Who knows, if Londoners could go and sleep down there whenever they wished, thousands might do so right now.

Two nights before the 177,000 peak of 27 September, one of several underground observers covered the scene at a Tube station with two typical platforms, a hundred yards long, five wide, with a painted line reserving the front yard for train passengers. There were three benches spaced out along the walls and two lavatories (buckets behind hessian screens), one for men and one for women, at opposite ends of each platform. At 7.10 p.m. there were about 300 people on each platform; another 200 in connecting passages, and well over 100 spreading up the emergency stairs – that is, about 1000 all together. The numbers of women and children were pretty well complete by 7.0 p.m., whereas the numbers of men went on increasing steadily between 7.0 and 8.0, and at a slower rate between 8.0 and 9.0 (to nearly a third of the whole population).

These final 9.0 p.m. proportions – of about 50% women, 30% men and 20% children – were found to be pretty constant in all the Tube shelters studied, and remained so throughout the winter.

By seven, the only free place is at the bottom of the emergency stairs. The last arrivals, at just after eight, sleep at the very top of the emergency stairs underneath a glass roof, with no protection at all.

This top site is fully used every night, although it must be more dangerous than almost anywhere else in town. Many undergrounders come to believe in the place, its ambience, as protection almost sacrosanct in quality. Nor do large casualties, deeper down, have the least measurable effect on Tube confidence – 20 killed at Marble Arch, 17 September; during October four stations struck inside three nights, including Balham, with 600 casualties; in January, 117 one night at Bank station in the city.

Most of the believers are working-class people from the neighbouring streets, though a few have trekked as far as from the East End. The men wear normal working clothes and so do most of the older women – many of these with very old and laddered stockings and curlers. A large number of the younger women wear trousers or siren suits. The men take off their coats to sleep; the women usually lie down as they are.

Most of the people in the station are 'regulars'; they settle on the same bit of platform every night and know the 'neighbours'. Late-comers are greeted by those perched on the stairs with bulletins about current accommodation: 'Hallo, Joe, you won't get anywhere tonight. They've been down there since two this afternoon.' Someone stops and says to a group on the stairs: 'Excuse me, but weren't you the people who lent me a cigarette last night . . ." – and insists on repaying it.

By seven o'clock, the 'regulars' are settled in. Some of the children are tucked up with a blanket, sometimes an eiderdown. Many people are eating – sandwiches, chocolate and fruit – and are drinking tea from Thermoses, some beer. All are surrounded by blankets, pillows, large bags and packages. There is a babel of talking, echoing off the arched roof, but not much activity or laughter. Mostly the talk is about last night's raid. 'Fairly bad.' One man, who has spent the night in Lewisham, holds his audience enthralled by tales of the extent of the damage there.

Of amusements, or indeed any sort of occupation, there is almost nil. On one platform two groups are playing some sort of a card game, and a few people are looking through magazines in a half-hearted sort of way; hardly anyone has an actual book to read. The majority just sit doing nothing, staring into space, with an occasional exchange of remarks, or little bursts of conversation at infrequent intervals. It seems remarkable that so much noise could be generated by so few people actually conversing at any one time; the acoustics of the curved roof and the long narrow shape have something to do with it.

At intervals during the night, a man will go upstairs for a breath of air. As he comes down again, he will tell what is going on. This will give rise to prolonged discussion. One man, for instance, says: 'There's a fire over the way'; and various possible localities are then suggested.

Between 7.0 and 8.0, there is considerable grumbling about the 'booking' system. Those left in the passages and stairs are annoyed that others should have taken the best places by 3.0 in the afternoon – well before this particular station was officially 'open', at 4.0 p.m. Accusations of 'bribery' and 'buying places' are bandied about – and indeed, enterprising youths have been wangling their way on to the platforms early and 'selling' places for anything up to 2/6d each.

One grumbler is a woman who can find nowhere but the stairs, and has parked herself in the middle of them; she then tells everybody, in a loud voice, the way she has been treated. Nobody listens, and after a while she lapses into silence. Then a policeman appears and tells her to move to the side; this starts her off:

'I'm a ratepayer, I pay my rates as good as anyone else, I can sit anywhere I like!'

The policeman tells her to move again, and again she starts to rant. The policeman just looks at her without saying a word; she moves.

'I'm a ratepayer!' – this is often brought out as a trump card. Later in the night this same argument is used by another man who wants to sleep on the stairs in a position that would stop anyone passing.

There is another woman, at the bottom of the stairs, who grumbles every time anybody goes up to the lavatory at the top. These mostly apologize for disturbing her. But at about half past one, when most people are sleeping, her voice is suddenly heard complaining about having to move because '78 people want to go to the lavatory'. At this, a woman leans over the stairs and shouts:

'Can't you shut up, you bleeding little hypocrite?'

'I want to go to sleep and these people keep on going to the lavatory!' retorts the first woman.

'Yes, and so do I!' yells the second one. 'Why don't you move if you can't sleep there?'

The first woman does not reply, and nothing more is heard of her that night.

The people are all 'very obedient and friendly towards the officials and practially everybody gives a 1d or 2d towards the fund for the porters who are to clear up in the morning.' Almost no bitterness or resentment towards the enemy is expressed. References to the Germans are of a good-humoured, joking kind:

'You know, I don't think Hitler likes us much!'

'Bet we sleep better than old Hitler!' and so on...

About five, still dark, the guns still firing, people begin to leave. Most of the first leavers are men going to work; by 5.15 whole families are leaving, most of them for home to get a few hours' sleep before starting the day. The lifts are not working and all leave by the emergency stairs. At the top they hear the guns and decide to go back. But more are coming up behind, there is no room for more than one line of traffic on the stairs as some are still sleeping on the steps, blocking the way. The result is a solid mass of people, some trying to get out, others trying to get back in again. The crush is terrible but orderly. Soon the all-clear goes. From then on there is a steady stream leaving.

By six, twenty minutes after the all-clear, there are not more than six groups left on the platform. At seven o'clock, the platform is *completely* empty. All the abandoned mess of papers and fruit has been cleared up by the porters. The day trains roll in as usual.

The big shelter scene was never anything like what had been expected. Usually, not always, it was as orderly as everyday life, or more so.

(11) 'Tilbury'

'Kick it down!'

'Yes, kick it down! Kick down the bloody gate! Off with it!'

The sirens had gone: the building was heavily locked. Men were battering at the doors with boots and fists; women crying and sobbing:

'What shall we do? What shall we do...?'

'They'll be here before we get a chance...'

'Kick it down...!'

'We'll all be killed...!' – the shriek of a forty-year-old woman sounded above the rest: 'We'll be killed, I know.'

The Tubes were not the only mass shelters which 'the people' took over, more or less quietly. All over London a minority were to cram themselves into the basements or cellars of any really solid-looking building with idle space. The uproar recorded above was outside the largest, taken over early on; Tilbury Shelter.

Nobody seemed to know why it was called Tilbury Shelter, for it was miles from Tilbury Docks. It lay in the heart of one of the poorest areas of London, among the blitzed streets between Commercial Road and Cable Street – a couple of hundred yards from the junction where Whitechapel Road merges with Aldgate. It was a massive structure, ten storeys high, with huge, thick walls of brick and concrete, giving an impression of vast solidity. The basement to be commandeered as a shelter had been used in peacetime as a warehouse and goods depot. It was still full of crates of margarine, bales of newspaper, etc., when the shelterers first surged in, in unstoppable thousands. The general features of the place as described then:

> . . . first impression was of a dim, cavernous immensity. The roof is made of metal girders, held up by rows of arches, old and solid . . . giving a somewhat church-like atmosphere. The hall is oblong in shape, and covers many acres . . . so huge and dim, the end seems out of sight . . . Between the lines of archways are wooden platforms, raised about 4 feet from the ground, and stretching the whole length of the shelter. Between them are wide gang-ways, paved with brick and earth trodden so hard that one might imagine it was stone. Round three sides of the interior runs a narrow railway-track, in almost total darkness, for here there are no lights: and here the earthen floor is dark and rough. The entrance to this vast, dim, cathedral-like structure is narrow and insignificant – just a break in the street wall that could easily be missed by a passer-by. But once through this gap, one finds oneself in a large stone courtyard, sloping away in two directions down into the earth . . . There are only two small doors into the shelter, one of which can be locked on occasion, the other manned by police . . .

It was into this building that some ten thousand people crowded, night after night, from the second week of September onwards. At first, before the authorities realized what had hit them, conditions were described as 'indescribably awful'. One of an M-O team who visited the place on 14 September gave a personal impression:

> . . . a dense block of people, nothing else. By 7.30 p.m. every bit of floor-space taken up. Deckchairs, blankets, stools, seats, pillows . . . people lying everywhere, on the railway-track, among the margarine-crates, everywhere. The floor was awash with urine . . . only two lavatories for 5000 women, none for men . . . overcome by the smell. People are sleeping on piles of rubbish . . . the passages loaded with filth. Lights dim, or non-existent . . . they sit, in darkness, head of one against the feet of the next . . . there is no room to move and hardly any to stretch. Some horses are still stabled there, and their mess mingles with that of the humans . . .

Ten thousand people had come together without ties of friendship or of economics, with no plans at all as to what they meant to do. They found themselves, literally overnight, inhabitants of a vague, twilight town of strangers. At first there were no rules, rewards or penalties, no hierarchy of command, no London Transport porters. Almost immediately, 'laws' of a rough-and-ready sort began to emerge – laws enforced not by police and wardens (who at first proved helpless in the face of such multitudes), but generated by the shelterers themselves. One squabble and its solution set a precedent for the next, until, by nobody's deliberate edict, a system of rough justice was evolved; and adhered to. By the beginning of October, one could summarize this *ad hoc* code of behaviour:

'*How Order is Kept*' – Laws and Rules
There are two sets of restrictions.

a) The Port of London Authority by-laws, which restrict trading without permission, forbid sales of the *Daily Worker*, and give PLA police a certain amount of authority.

b) Unwritten laws, which are much more numerous, respected and effective. They have been evolved out of necessity. Principally they are:

Every passage-way to be kept clear. Gangway to be left through every arch. No one may remain stationary on these paths; no one may sleep on them; groups may not form on them.

It is an unpardonable crime to tread on anyone's blanket.

Music, singing, etc. to be stopped at 10 o'clock in the main part and at 12 o'clock in the space by the entrance at the back.

'Places' are respected. If a person has reserved a place, then he can go away and leave it, knowing that it will be there when he returns, provided he has covered it with a blanket.

There is a limit to the amount of space one person may reserve. Everyone can see what 'too much' is, and people with loads of blankets to spread around are kept under.

Infringement of any unwritten laws about 'places' invariably arouses an upsurge of fury, not only from the victim, but from bystanders also. Cries of 'Bugger off!' . . . 'Get yer f . . ing arse outer 'ere!' greet such infringements.

Concurrently with these more or less spontaneous developments in the direction of law and order, there was inevitably an attempt by the authorities to impose some system for running the shelter. The first step in this direction was the provision of basic amenities. Lavatories were installed, and teams of morning cleaners organized. Wardens were seconded from neighbouring posts and put on duty in Tilbury – not many of them, because the posts themselves were hard pressed for personnel, with the bombing continuing night after night. The PLA police were set to patrolling the place, and the Metropolitan Police were also given duties in the shelter – to begin with police had been operating under great difficulties, since Tilbury was not a recognized public shelter, but still classified as 'private premises', so that the police had (in strict law) no right to interfere inside. All they could legally do was try to control the crowds outside the gates. When the shelter opened in the afternoons, they would link hands to try to stem the rush down the slope; but, as one of them remarked, it was like 'holding back a stampede of buffaloes'.

More in evidence than either police or wardens, were the 'shelter marshals' – often self-appointed – and soldiers intermittently brought in to help keep order.

As late as 21 October this 'little Republic' atmosphere was still in evidence:

... The police have not much authority and the shelter as a whole presents a huge opportunity for 'Little Hitlering'. The police and the wardens themselves are fairly reasonable, but the soldiers and shelter marshals – offices requiring far fewer qualifications – take advantage of their authority to bully the rest of the people.

There are a great many shelter marshals. One, aged 50, shouts loudly all the time the most unnecessary instructions, such as:

'Don't tread on that woman's toes!'

'Don't keep bashing into each other!'

He is continually having arguments with overwrought women, he himself being highly strung and easily rattled.

A young soldier has the most domineering attitude. He wanders around the shelter all through the night and early morning shouting:

'Pass along there!' and:

'Break it up, you blighters!'

There is another form of order-keeping – the public protest. If a torch is raised too high, or if someone starts shouting in the middle of the night, or if anyone smokes in the main shelter, a chorus of irate voices yells a protest – 'Keep that light down!', 'Pipe down!' etc.

Quarrels and arguments before ten o'clock may be dealt with by police, wardens, etc. After ten, the would-be sleepers take them in hand.

As described, the roof was supported by rows of giant arches, running the whole length of Tilbury. The area within each of these arches came to be referred to as a 'bay'. Within three weeks people had sorted themselves out into these bays, and felt themselves to belong there. Each bay held between 50–100 people; sometimes the nucleus of the group was a family or group of families, and their friends. But this was by no means always the case. Many of the bay groups started off as strangers; but by the middle of October they all showed a high degree of cohesion. 'Bay loyalty' was strong; if a newcomer turned up and began dumping his things on an apparently empty space within the bay, he would be driven off by a chorus of noisy protest.

Bay 7 gives an example of bay organization. The platform within the arch holds about 40. The shelter marshal is Mrs X, a little wiry woman, 50. She is tireless in her energy and enthusiasm for keeping the bay in order.

'Every morning I take the beds,' she says, 'everybody who leaves them, you know, and I have some of the kiddies help me, and I take them home with me and put them in my front room where it's nice and dry. When the sun shines I hang them out in the garden, make them nice and fresh you know. Then after I've had my dinner I brings them back and puts them out all nice where they belong. Then they find them all ready for them when they come in from work, they don't have to wait about here, I see to it all for 'em. We all work you see, all us here.'

Mrs X prides herself on keeping the peace in her bay. 'We ain't had no fights here, not on my platform, not since I been here. We all has our place and we keeps to it.' 'Fighting' and 'sex' are the two things which dramatize the monotony of Tilbury nights; neither of them are tolerated in the more 'official' Tube underworld. Here they are greatly exaggerated in chatter. In any one bay, people would say it only occurred elsewhere. Prolonged observation of the shelter night after night revealed very little of either. There was a certain amount of necking among young couples – usually broken up smartly by relatives or neighbouring groups; and occasionally one would glimpse a couple engaged in intercourse in the dark area of the old railway track. But this was the exception. So, too, was outright fighting, despite frequent assertions to the contrary. A typical case was that of a young girl shouting and screaming at her mother, and waving her fists, while noisy threats were exchanged: in the end, the two were separated by force and led away from each other, struggling and screaming. Another time, a wife wanted her husband to sit down – he wanted to walk about. She became very excited, and, as a circle of spectators formed round them, she threw herself upon him, biting his ear and tearing out his hair. He smacked her face and threw her to the ground . . . at this point the wardens intervened and separated them.

Occasionally, in one bay or another, some sort of spontaneous revelry would arise. This was always more of a popular success than were the occasional organized entertainments.

In the middle arch, on Friday, a girl played an accordion, while men danced burlesque dances round her. She was a tall, pale girl with long straggly hair, and played effortlessly with a vacant face. The group of spectators was constantly being broken up by the wardens.

After a while a man in the arch took hold of a woman's hat and put it on back to front. Then he sidled up to his mate, Mae West-wise, and made some kind of wisecrack. The crowd roared. Gradually the two worked up some spontaneous cross-talk. The girl got on to a box of lard and played louder. The crowd joined in lustily . . .

On another occasion a young coster was playing the accordion. He played well – with fire. A coster girl, about 20, sang a gypsy melody in a clear, high, plaintive voice . . . Another archway was playing 'Knees up Mother Brown'.

Another night, a crowd of drunken seamen made a chain and ran around the aisles, chanting sea songs. All efforts on the part of the wardens to break them up proved fruitless.

But such scenes as these had to be sought-out for the record. Observations of randomly selected groups and individuals gave a far more apathetic picture. Typical of one individual:

4 p.m. A poor working-class woman makes her way slowly down the aisle, carrying a heavy bundle of bedding tied up with string, and fish-basket filled to overflowing. Though only middle-aged, her hair is grey and her face wrinkled. She is wearing a dark, much-worn coat with moth-eaten fur collar. It has no buttons, and she clutches at it with the hand holding the fish basket to keep it together. She has a pair of very worn black strap shoes, trodden down at the heel, and brown woollen stockings wrinkling round her ankles. She walks slowly but purposefully down the aisle till she gets to the platform; then begins looking around:

'Got a place, Evie?' calls out a woman of about the same age, who is already established among her blankets on the platform.

'Dorrie come to git me one; have you seen Dorrie?'

'What, Dorrie with Carsons?'

'Nay, not her. Dorrie what comes up here along of us.'

'Oh, her. No, I ain't seen her. No, I ain't seen Dorrie. Jock, here, Jock. Have you seen Dorrie?'

'What? (to Evie) You looking for Dorrie?'

'Yes, have you seen her?'

'No, I ain't seen her.'

The woman continues on her way, glancing from side to side. Suddenly she stops, looks eagerly at the far corner of the platform, makes a movement as if to put down her bundles, then tightens her hold on them, takes three hurried steps forward, hesitates, turns back on her tracks and walks towards the steps up to the platform. She climbs them, breathing heavily, then stumbles along hurriedly, picking her way among bedding and reclining people, to the far corner. There, against a wooden partition at the end of the platform, a grey rug is folded lengthways. She dumps her parcels down just in front of it and starts to unfold it to its full width; finds her baggage in the way, picks it up again and moves it further away. She takes well over a minute unfolding the rug, breathing hard all the time. Having done this, she sits down on it heavily, leans against the partition, and stares in front of her for several minutes without doing anything. Then she gets up once more, and begins to untie the string of her bundle. It falls apart – a greenish-grey eiderdown, two overcoats and a pillow. She picks up one of the overcoats and looks at the floor consideringly; then seems to change her mind, puts it down, and sits down again, leaving the things as they are.

5 p.m. Still sitting there surrounded by the things she has brought. Stares expressionlessly at first one lot of people and then another. After some minutes a man, about 45, picks his way towards her across the by now crowded platform. She looks at him as he approaches, and as he lowers himself on to the rug beside her. Neither of them says anything. He sits for a minute doing nothing, then slowly begins to loosen the laces of his boots. He takes a very long time doing this; then pauses in the middle of it:

'What you got there, Mum?'

She does not answer, but reaches for the fish-basket. She burrows her hand under what appears to be some kind of woollen garment, and draws out a newspaper parcel and a bottle of cold tea; she digs again and fishes out a china teacup. The man has undone the parcel meantime and is already eating a thick cheese sandwich. She pours out some tea, hands it to him, and he drinks it without speaking. He puts down the sandwich half-eaten and goes on loosening his boots.

So the record slowly unrolls – the sandwiches scattered over the rug by 6.16 p.m. to be nibbled sporadically with the help of a newcomer, a ginger-haired girl of 20. 6.30-ish two middle-aged women drop by; 7 p.m. both man and woman have a nap under eiderdown and overcoats; girl going with friend to canteen, returning soon after 9, when she and the woman, awoken, eat sandwich remains while the man sleeps on. By 10 p.m. all three asleep, settled down.

On the basis of many such long observations, a general impression recorded for Tilbury as a whole; from mid-September into November:

The first thing that strikes one is the extreme vagueness and lack of plan that characterizes the activities of individuals here. There is an enormous amount of simply sitting about with no occupation of any kind. Pastimes such as reading the paper tend to be very half-hearted, the person continually fidgeting, looking about him, and so on. Very few people come equipped with any plan for how to spend their time. The main tendency among younger people appears to be a vague wish to run across some acquaintance or pleasant stranger. But even among them, any positive attempt to bring this about is the exception; mostly, they just sit around waiting for something to happen. Older people seem to feel little need for entertainment or conversation of any kind.

This was, in brief, a huge agglomeration of families with individual acquaintances, each doing their thing as opportunity allowed, within an improvised order which had little social content or cohesion. Summarizing these processes, randomly selected families were observed from arrival (often as early as 3 p.m.) to leaving next morning. The only major difference between the sexes was, once more, less disturbed sleeping for the men generally, while younger people not unexpectedly tended to be more active than their elders, especially as regards somewhat 'aimless' moving about.

Here is one 'night's' record of continuous observations on 20 individuals. To simplify, the main hourly activity is rated from 1, the major preoccupation of the 20, to as low as 7 where only a very small minority were so occupied; a dash means no one was pursuing the activity in question.

TILBURY SHELTER: MAIN ACTIVITY RATING (1–7 SCALE)

	NECESSARY ACTS			LEISURE			
Time	*preparing sleep place*	*sleeping*	*eating*	*Knitting, reading & other personal acts*	*Talk and contact*	*'Doing nothing'*	*Moving about*
p.m.							
3–4	3	2 (naps)	–	5	4	1	5
5–6	4	–	2	6	5	1	3
7–8	7	1	6	5	4	3	2
9–10	–	1	–	4	–	2	3
11–12	–	1	–	–	–	2	3
a.m.							
1–2	–	1	–	–	2	3	4
3–4	3	1	–	–	2	–	–
5–6	2	1	3	–	–	–	4
6–7	4	1	3	–	–	–	2 (leaving)
7–8	4	2	3	–	–	–	1 (left)

Although sleep dominated the Tilbury landscape over a twelve-hour span, at no time were over two-thirds of the shelterers sleeping at once. At 1 a.m. there was a sub-peak of revived conversation and mobility. After about 5.30 a.m. most were packing or leaving, but a minority – roughly 12% for the whole shelter – slept on after 7 a.m.

(12) The homeless

Those 'vast minorities' of Tilbury and the Tubes, officially almost ignored, included many shades of apathy, anxiety and distress, towns within a city, where life was stripped down to near the bare bones of group and self-survival – a sandwich, a blanket, cover, a little talk. Tens of thousands of Londoners found the loss of private identity worth it, where they did not, in part at least, actually prefer sharing fear and discomfort. Some went to these great places because they could not feel secure in their homes; and they were often right. But it is by no means safe to suppose – as many since have – that those nightly inhabitants of Anonopolis were solely, even mostly, extra-nervous potential raw material in the great panic that never came.

Each citizen, in those dark times, faced inner tensions, however obscurely, in his or her own way, whether deep under Liverpool Street, in a surface shelter on Smithy Street, Stepney, dug into their own little Hampstead Anderson, or huddled under their bedclothes off the Fulham Road. Millions – at least 90% of those who stayed in London – managed, after a sticky start, to develop into a vast, largely self-disciplined, many-layered working body. It is this enormous diversity which continually frustrates generalization about the civilian mass.

A great many of those who took to the underground way of life at this time had homes still standing. Mother and kids went back there for the day. Others had no other place to go. Homelessness was nearly as much a state of mind as of fact. Which is not to say that homelessness was not very important in its own right. With over a million London homes eventually damaged, one in five temporarily uninhabitable or worse – two in five in Stepney – the physical problem was gigantic. So were the administrative problems, though the enormity of London greatly helped in patching up holes, just as the concentration of skilled leadership there quickened responses to breakdowns.

Terence O'Brien, the official historian of Civil Defence back in 1955, looking at what happened during 1940, considered that, after the difficulties experienced by the blacked-out ARP and other city services, 'the second most significant unexpected feature of these attacks was the relatively small loss of life'. In contrast, major and minor domestic damage was high, leading directly (as he says) to 'problems of quite unexpected character and magnitude in the sphere which became officially called post-raid services!'

The most immediately distressing of these post-raid problems was the plight of the more or less homeless. By the end of September some 25,000 of these in the London region were being catered for at organized Rest Centres, most of them under intensely crowded conditions within the London County Council limits. Richard Titmuss in his volume on the social services at war, gave this a full chapter – 'The Challenge of London's Homeless' – which traced out the official record. As he shows 'a philosophy of life, cool, detached and secure' failed to contemplate the possibility that such things as clothing, rough shelter, soup and margarine might have to be provided. That was the 'poor law mentality' of peacetime. It was *inconceivable*, according to this philosophy, that the accident of war, 'even with the bombers thrown in', would require a drastically new approach. In particular, the peacetime traditions of the civil service led to the poor law authorities being initially responsible for part of the homeless problem. The supply of food was considered as one thing, of temporary shelter another, while the need for rehousing bombed-out families was not clearly foreseen – 'a problem to become the most critical of all the social consequences of air attack'.[8]

We would not, for once, agree with Titmuss in that last quotation, in so far as he extended it to outside Greater London. In the provinces simple human distress was surely most critical. This, and the whole question of how relief machinery worked in practice, will be a main concern of following chapters.

By November the Rest Centres in most of London were becoming well organized – heavenly as compared with the places so classified in other blitzed towns. We must be careful not to over-emphasize the administrative defects, in London where they were less, or elsewhere where they tended to be much greater. An astonishingly large number of civilians clung to the bright side. A fifty-year-old worker in a small London factory making ornamental buttons was asked, after he had been bombed-out twice, if he and his wife would like to be evacuated as non-essential personnel.

'What, and miss all this?' he exclaimed, '(not) for all the gold in China! There's never been nothing like it! Never! And never will be again.'

London's monopoly of heavy bombing ended on the night of 13 November, when a mere 25 German bombers came over, letting go a derisory 28 tons of HE – against a total of 13,651 tons (and 12,586 incendiary canisters) to that date, giving an average 201 tons (and 182 canisters) per night since 7 September. Later, on 19 March 1941, 479 bombers hit London and on 18 April an all time record 685. But the region now went for weeks with little activity, starting from 14 November when 449 planes switched to Coventry.

In all, London received 18,800 tons of explosive in its 1940–41 blitz. Of this, 5200 tons fell in the second, irregular period, until 16 May when the whole blitz abruptly stopped after a last attack on London (711 tons) and on Birmingham (160). The 13,000 tons plus that fell before mid-November exceeded, in gross, all those falling on all places blitzed in the following period, London and the provinces included.

Thus London had the doubtful privilege of providing an intensive training course in bomb reception, on a scale and of a rhythm unmatched elsewhere. Many London trainees learned faster than their leaders. They had to, in order to keep themselves or their spouses earning money and sleeping. But everyone shared in the process willy nilly. It was of the greatest help that the assault was so predictable and broadly similar from night to night. For the recipient, the only difference between one night and the next was the amount of *localized* uproar, and this mattered less and less as time went on, not at all in the sound-proofed underworlds of Tube or Tilbury and the like. One of our most regular war diarists put it well, in another way, when she wrote rhetorically addressing Hitler:

We are a little frightened, we who have been happy,
We are not frightened enough to become what you want.
We set our will against yours, the will of London,
If you kill us, we only die.

Some day the story will end, the book be shut for ever,
Sleep will be sweet again.

The much less regular raids of Christmas time, full winter, did something to disturb these routines, though for most night-habits had become settled, regardless not only of bombs but of warnings and other external stimuli. This was more so in the crowded boroughs, where many felt more vulnerable than in the outer suburbs with houses further from each other. Compare, for example, the typically middle-class scatter of Harrow and Hendon, with 25,000 acres, to Stepney and Shoreditch with 2400. The former received only 20% more bombs than the latter, which had ten times the bomb intensity per acre in one blitz month.[9]

Sporadic harsh experience was, on the whole, much harder to take than continual unpleasantness. In this, the Luftwaffe helped Londoners by bombing so steadily at the start, and never giving up for too long later on. Had they chosen a different pattern this story would surely have had to be constructed differently. Continuity of experience also helped at the institutional level, where government and administration had to learn from the new experience – and learn fast under steady pressures. Although the lessons to be learned strikingly failed to percolate outwards to the provinces with speed and force, at the centre there was a whirlpool of activity, rotating about Downing Street and the House of Commons, Whitehall, Buckingham Palace, Scotland Yard, the LCC headquarters just across the Thames, between them holding at least nine-tenths of the nation's top decision-making machinery.

In October 1940 radical Miss Ellen Wilkinson ('Red Ellen') was appointed Joint Parliamentary Secretary to the Minister of Home Security (now and all through the blitz Herbert Morrison, previously LCC leader). She had special responsibility for shelter matters – a quiet recognition that the official shelter policy had hitherto been inadequate. But the size and complexity of the machine and the long time lags in wartime construction, ensured that progress was not rapid. In February 1941 one Londoner in five remained technically 'unprotected' by any kind of officially provided shelter.

At the end of 1940, the government had developed plans to dig enormous tunnels under existing Tube facilities, including under ten big station platforms, to give added deep shelter capacity for 70,000. The first cost estimate of £1.5 million was far exceeded, through miscalculations and labour difficulties which also delayed opening until long after the blitz was past. Shelter improvements of other kinds were put in hand, especially in the neglected direction of minor comforts: heat, light, bunks, toilets, etc.

Had the London blitz long continued, a major shift in shelter attitudes could have resulted. Already as spring brought less regular bombardment, Londoners commonly felt more looked after, less self-dependent, more comfortable generally. That vague miasma called 'public opinion' moved a long way through the winter. The contemptuous hatred which so many surface folk expressed towards deep shelterers was practically forgotten. Official distaste for 'timorous troglodytes' had been correspondingly outdated. By the spring of 1941 it was difficult to remember the violence of early prejudice against Tube crowds the previous autumn, when they had been abused *en masse*, as dirty, smelly, lousy, diseased, cowardly, foreign and (especially) work-shy. A venomous campaign, encouraged in the press, attacked the so-called 'Tube Cuthberts', young men of military age supposedly hiding down below. Not often were wartime Britons so positive as some were on this issue:

Greengrocer, aged 45: 'It's perfectly disgraceful I think. Why, no decent man would ever dream of doing such a thing! Women and children have got to come first, always.'

Balloon Barrage Operator, aged 35: 'It's disgraceful! All those women and kiddies, they need all the space and air they can get ... They ought to be in one of the Services these young fellers, that'd teach them a thing or two ...'

Housekeeper, aged 35: 'I think it's absolutely disgusting! I'd like to see one of my brothers (in the Services) getting his hands on them! I can't understand them being willing to do a thing like that, I know I wouldn't if I were a man ...'

Serviceman, aged 35: 'They ought to 'a joined up long ago, if they'd any decency in 'em ...'

Baker's Roundsman, aged 40: 'I'd never have believed there'd 'a bin so many to behave so disgusting! It's a disgrace to the whole country!'

In real life, below the verbiage, there cowered no category of Cuthbert – a term originating in 1917 to describe an officer shirking military service. Among the few young men down below, nearly all were waiting for call up, as military service was now by conscription. Some had been living alone on top in lodgings or with their parents evacuated; others were homeless. Quite a few turned out to be servicemen on leave in mufti, accompanying their families.

Tilbury and the Tube became respectable. Special trains had long been running down into Kent, taking a new sort of commuter, the cave men and women of Chislehurst, whose home from home developed as a private, paying enterprise with its own electricity, church, barber and entertainment, How distant that underworld seems today, as seen by a woman who tried it out in November 1940:

Chislehurst Caves

We have had bad nights and Mrs S got very nervy so on Sunday 10th we packed up bedding and we took the car to Chislehurst Caves. I spent the night there. We were late and didn't get there till 5.45 – there wasn't a spare yard to put my camp bed in, so we sat down quite near the church, on our bundles and waited. I went to the service, there were between 3 and 4 hundred there, it looked as if it were meant for service – even a natural dome rising above the improvised altar – there was a small organ or was it a big harmonium, anyway it gave out a good sound. Strangely enough the first hymn was – Rock of Ages, cleft cleft rocks – what a memory that hymn will be – always when I hear it I shall see before me that strange scene, hear that grand singing.

The service which began at 6.30 only lasted about half an hour, and then we trooped out and the door was shut. Right through the caves were families gathered round in groups – or resting on their beds, with candles stuck in the ledges of the rocks – reading, playing cards and a continual hum of voices was heard – then the canteen close to the church was opened and we saw long queues of folk all holding jugs, teapots or cans waiting to be filled by tea (milked and sugared). I think cake and biscuits were also sold. This went on till 10, and then we put out our sleeping things.

We were told to go to the inner caves: but they had been filled by regular visitors – who had commandeered positions weeks before –

some had taken possession of cut out rooms, and curtains were fixed in front and behind, whole families lived – and there were tables, cooking stoves, beds, chairs behind the curtains – bombed out families live there *permanently and the father goes to work and returns there and the mother goes out to shop and that is their home*. There were 8000 in the caves! When we got there a man near us constituted himself our helper and put us wise to several things and when I lay down, he and his wife tucked me well up.

The sanitary arrangements were good – parts partitioned off and were disinfected – but alas, one was too near us – a man's – and one would have thought that everyone would have finished by bed time but all night long the sound went on and on and on. I got to sleep at last – soundly – but at 4 a.m. everyone woke up and coughed for half an hour – then went to sleep again – the current of air was changed then I hear – 5 a.m. – the lights were full on and the canteen opened. But I wasn't disturbed, as there were not too many crowding up for tea. At 6, we got up and folded up our things – I walked down the cave and came across a big woman folding up a bed exactly like mine and told her it was the first I'd seen – she was putting it on her wooden bedstead, with a small chest of drawers. She said it was a good camp bed, but her husband was ruining it, as he weighed 17 stone.

Poor Mrs S was very upset by it all – 'I'm not a snob,' said she, 'but what a dreadful place.' And she preferred to be nervous at home. Her husband and I laughed: if she'd mixed with folk as we had, she wouldn't have been so upset. We arrived home at 7.30 and I am very glad to have had the experience.

6 ‘Coventration’

‘We could see no reason why the hostile bombing of London should not go on throughout the war.’

W. S. CHURCHILL, 1949

The heat came off London on Thursday night, 14 November 1940, after over two months of intensive bombing. It was turned on to middle-sized midland Coventry, with a quarter of a million inhabitants, best known for engineering industries and the fourteenth-century Cathedral. 449 German bombers struck, more than in any previous London night, 503 tons of HE fell – nearly twice the cumulative total on any one London borough during all the preceding month – plus 881 incendiary canisters. It thus became the biggest air-raid ever in the British Isles and easily the most concentrated. Next morning, nearly everyone there claimed to know someone among the 554 killed or 865 seriously hurt. This noticeably brought the experience ‘home more fiercely to everybody’. There was no gradual human conditioning. The word *blitz* was truly appropriate.

But human physical hurt was small, visually, as compared with the devastation: rubble, debris, broken glass, charred timber, a total mess concentrated impressively in roughly a hundred acres of the centre; but spread all over the rest of the city too. Two hospitals were badly hit, along with 21 major factories – the ostensible targets. A relatively objective Beaverbrook-press journalist, Hilde Marchant, happened to be staying in adjacent Birmingham that night, so she was able to reach the smaller city while it was still smoking. On the way she met lorries ‘packed with women and children’, sitting on bedding bundles, fleeing the area. Other people ‘leaned against the railings at the roadside, too exhausted to move’, as her later published account tells. She thought these people ‘did not yet understand that they had survived'. In the city centre, metal was still red hot. Only the police as a force seemed (to her) to keep their cool.

I was a bit further away, yet among the first outsiders to penetrate the police road-blocks. On Saturday night I also penetrated a considerable news blackout to tell about half of the Coventry truth on the BBC Home Service at 9 p.m., then the most prestigious spot in the day's radio. For by then rumour across Britain had so magnified 'coventration' that the authorities decided it best to have some sort of factual observer tell the tale. Among other things, I broadcast this:

The strangest sight of all was the Cathedral. At each end the bare frames of the great windows still have a kind of beauty without their glass; but in between them is an incredible chaos of bricks, pillars, girders, memorial tablets. And all through the town there's damage – rows and rows of houses, smashed in windows and leaking roofs. (Nearly one third of all homes had been made uninhabitable.)

As soon as darkness fell, the streets were silent. The people of Coventry had gone to shelter. For more than an hour I drove in my car picking a way through craters and glass, and during the whole time I saw no other private car. I think this is one of the weirdest experiences of my whole life, driving in a lonely, silent desolation and drizzling rain in that great industrial town. Then I met a gang of young men looking for a drink. I went with them because I wanted one, too. We must have visited half the pubs in Coventry before we found one.[1]

The human picture was, however, better picked up by the team of mass-observers who arrived on Friday afternoon, fresh from long London blitz experience and visits to many previous lesser bombings elsewhere. We also, by chance, had a team studying wartime savings working in pre-blitz Coventry, which gave valuable background and contacts.[2]

For the population as a whole, from the Town Clerk down, the raid seems to have come as a complete surprise – although oddly enough top Ministry of Home Security scientists had shortly before chosen the place as a typical target and made elaborate statistical projections on the effects to be expected.[3] Their report was secret, of course. There was no blitz anticipation. Preparations were weak.

Of 15,000 children eligible for official evacuation in September 1939, only just over 3000 had gone. At 20% that was one of the lowest figures in Britain – compare 70% for Newcastle, 69% Manchester, 61% Liverpool. Many had since returned. Most of the 13,000 still in town by 14 November were among the masses quitting the city as seen (though not reported publicly) by all of us outsiders next day.

Sudden, unexpected, near 'total' assault is most shocking. In Coventry this produced 'unprecedented dislocation and depression', compared with all we had hitherto observed elsewhere. Among the ordinary inhabitants: 'those who were even able to walk about, were left to themselves'. This phrase – 'left to themselves' – was to be echoed in many later blitz situations. The Mayor, Town Clerk, and all their officials faced dawn as pulverized as anyone else. They were barely able to rise to their enormous new responsibilities. Even elementary information services did not function. Few knew what was still operative or where it was. There was no leadership at any mass-level. This phenomenon had been masked or overcome in London, partly by the period allowed for gradual learning, more by mere structure and size. In and around the capital were other human arms or organizations ready to help. In Coventry suddenly, there was an aching nothingness. As we wrote, there and then (original italics):

There were more open signs of hysteria, terror, neurosis, observed than during the whole of the previous two months together in all areas. Women were seen to cry, to scream, to tremble all over, to faint in the street, to attack a fireman, and so on.

The overwhelmingly dominant feeling on Friday was the feeling of utter *helplessness*. The tremendous impact of the previous night had left people practically speechless in many cases. And it made them feel impotent. *There was no role for the civilian. Ordinary people had no idea what they should do*. And this helplessness and impotence only accelerated depression. On Friday (15th) evening, there were several signs of suppressed panic as darkness approached. In two cases people were seen fighting to get on to cars, which they thought would take them out into the country, though in fact, as the drivers insisted, the cars were going just up the road to garage. If there had been another attack, the effects in terms of human behaviour would have been much more striking and terrible.

If there had been another big attack ... It is difficult, today, to understand why there was not. Given the intent 'to bomb into submission', repetition would have seemed logical. But although two or three nights of blitz in succession were to become a commonplace in the provinces, this time the bombers switched back to London. The same lack of a psychologically meaningful plan for civilian bombing was to characterize Britain's own efforts on Germany.

Be that, for the moment, as it may, nearly everything in Coventry had broken down. Even for the determined, if a house was damaged alternative accommodation was almost non-existent. It got to be a major trouble to find a cup of tea and a bun. Hear the experience of one hungry mass-observer early that afternoon:

Tried to find a café (or canteen) for tea. None was open. Therefore asked five people – two of them keepers of small general shops, the other three private householders – if they would make tea. Of the three private householders, one said that he and his family were ready to leave for the country immediately, and that in any case they had no bread and – like everyone else in Coventry – no water. Another said that she and her husband were about to leave the city with the little food they had left; and the last, who turned out to be a lodger, said that his landlady had left during the afternoon leaving himself and three others with no bread and very little food of any kind. Of the two shopkeepers, one hadn't any food apart from tinned stuff, and the other had nothing but biscuits. Bought a pound of biscuits.

'Coventry is finished' and 'Coventry is dead' were key phrases of that fishless Friday. Many for a while showed no hope for the place. They could only survive as persons.

The minority who chose to survive *in situ* mostly went to some sort of shelter before darkness fell on Friday – thus the deserted streets. The shelters were far from adequate, by winter 1940 standards. One of the less bad was in a brewery yard. Mainly occupied by elderly folk, it had no bunks; no sanitation; candlelight. 'The stench was overpowering.' Other shelters, visited at random, were always very damp. Some were empty except for foetid floor-water. Plenty of brick surface shelters had received direct hits. There were no deep ones. A good deal of the housing in the centre was too crowded for garden Andersons.

A large proportion even of those who stayed in town expected a second big raid on Friday night. When it did not come, their first dread passed. Not that even near-panic was ever 'positive' in the sense of civil disobedience, open defeatism, even fierce complaint. Escape was simply the obvious solution from a place where the social services failed to function.

Nor was the recovery in morale due to any marked improvement in these or other services. We specifically recorded the view that recovery 'was not due to official activities', which continued to concentrate on the material debris, with the initial aid of 600, soon 1100 and then 1800 soldiers, clearing, dynamiting, demolishing. But out of the rubble began to grow local pride. A first catalyst was the scale of the disaster. No one had ever suffered more! It was a wonder to have endured at all! Associated with this, some credit was given to the AA defences, though nearly everyone regarded them (there were 32 heavy guns and 56 balloons) as inadequate and there was abundant chatter that they had run out of ammunition soon after midnight. Noticeable too was a complete absence of scape-goating. When the water ran out at 2 a.m., no one could blame the fire-services – pouring in from all over central England – for failing to cope, though many regarded them as in some way inadequate.

Whatever the errors, military or civilian, these could be excused, just like one's own lapses of judgement. There was thus a recognition that all were equally vulnerable. In an unprecedented situation, unprecedented responses were to be expected; and excused even if anti-social. Yet the degree of anti-social action was extremely low. We have no record of rape, looting, or even plain verbal nastiness in Coventry; nothing of the sort civilian populations in Europe were later to expect from uniformed, disciplined allied occupying forces (as from others before them).

An 'out of the blue' assault was so unexpected by the citizen as to make it seem reasonable that everybody else was unprepared too. This attitude was, in varying degrees, to characterize popular reactions in all provincial blitzes. Oddly enough, however, this diagnosis of total vulnerability had not been foreseen by Britain's higher leadership, which had never conceived that key people might suffer just as much as those for whom they were responsible; and that in consequence, they might lose not only their capacity to react but, at least temporarily, even the urge to do more than look after themselves. Yet this was a major cause for things going wrong, or at the best not going right, in blitz after blitz throughout the provinces. Coventry was the first example.

Rumour provided another liberating outlet for those who lacked clear leadership as well as sufficient hard information. Rumour blossomed upon the ruins of Coventry. One of the more fearsome, first heard from a priest, told of difficulties at the funeral of raid victims 'because blood was dripping from the coffins'. Another, equally untrue, had a German bomber crashing on top of the Co-op shop. A third had a man shot while signalling with a night light. 'There were signals. No doubt about that,' said a woman voicing one of many rumours which attributed the whole disaster to guidance from the ground. On the same theme:

Woman: 'There was a swastika in the sky before the raid – not long before the bombs began to drop. To warn fifth-columnists to clear out.'

Man: 'How do you mean, there was a swastika in the sky?'

Woman: 'Made in smoke. I've seen aeroplanes make question marks myself.'

The compression of damage in the city centre struck many as especially impressive. Here and later, where social nuclei like cathedral and shopping centre were destroyed, there was a powerful dual impact: a first sense of drastic loss, closely followed by a quite passionate interest, growing readily to pride. Again, London's size and scatter masked the effect; while many of London's oldest focal buildings never were destroyed (St Paul's, Westminster Abbey, Buckingham Palace). In a place like Coventry, the lot could go in one night. This struck at the heart of the community. But that did not have to mean that the heart could not be revitalized later. Meanwhile, the deadly effect would fascinate, even please.

So it was that after 24 hours of peace, things looked a bit brighter by 9 a.m. on the Saturday morning, when our record reads:

Coventry people were looking calmer and more purposeful. The central area was thronged with sightseers. The roads leading from the city were equally thronged, however, with pedestrians carrying suitcases, carrier-bags, and bundles, and motors filled at the back with bedding and baggage.

All one could get for breakfast was biscuits and cheese and milk in a little shop-cum-refreshment room. The elderly shopkeeper remarked to another patron in between serving him and someone else that her husband was 'under that lot' – meaning the wreckage of a near-by surface shelter which had suffered a direct hit and under which lay many dead bodies. She introduced this information *quite casually*, without any apparent emotion and then went on with her work – which consisted to a great extent of telling people that she hadn't any bread.

One of us managed, later that Saturday, to get on a crowded bus to Kenilworth, taking that as a typical outlying hamlet. A quarter of the female passengers had cases or bundles for self-evacuation. Most of the more numerous males had been seeing about their jobs and were now returning to evacuated families. At Kenilworth, four Rest Centres were coping with other fragments of the overflow. The Parochial Hall held 93, sleeping on beds and palliasses donated voluntarily by local soldiers, 57 occupied the Abbey Hotel ballroom, 80 the Wesleyan Chapel, others St John's School. All available rooms in private houses were taken. The surrounding villages were crammed. Local residents were full of complaints about the invaders, shortages in the shops, and much else.

Back in town, the two Rest Centres functioning (out of 15) in a system supposed to succour all the homeless, were nearly empty for the weekend despite the thousands whose houses had been destroyed. The centres had no shelters. A report on these:

St Thomas's rest-centre comprised two rooms – a general room and a dormitory. The general room was about 24 feet in width and 40 feet in length. There was a good fire, and round it four easy chairs, a sofa. Half a dozen hard chairs were scattered about the room. There were three women and a man, all middle-aged, and two children in the room, sitting round the fire drinking tea served by the Vicar. In the dormitory there were a dozen beds. The Vicar took us into the kitchen to see the supplies – bread sent by the Ministry of Health, butter and bully-beef. Water for tea was boiled on the fire in the general room.

At St Barbara's there was one room only. Eight three-tier bunks were standing and about 50 were piled unerected in a corner. The *four women, four children and one man* in this centre were all better dressed than those at St Thomas's. The women were unpacking bedding from suitcases. They had all been sent there because of unexploded bombs [i.e. were not strictly homeless].

Left alone, Coventry people edged slowly forward towards something nearer normality. Ten days after the raids, new words were in the air expressing growing dissatisfaction with the powers-that-be-not, the continuingly confused utility services, general social inconvenience beyond the obviously unavoidable, lack of simple comforts (like cigarettes). Transport, for example, remained chaotic. Some citizens were becoming openly critical of what could now be seen as the crawling rate of social recovery. One man put the growing view that 'the authorities are making the most of the raid . . . they see themselves as martyrs with a job that will last for the duration'.

'Crawl' was apt. It proved a slow and also a timid process, loaded with civic anxiety. Many at the top and below shared the attitude of a middle-aged factory worker: 'I expect they left the (Cathedral) tower as a landmark for future engagements. When we've got things built up a bit they'll be here again. They won't let you rebuild.' Or a 35-year-old curate (Church of England): 'The plane over this morning was taking photographs, I expect. It was just the same for a fortnight before the big raid. They were over every day, but nothing much happened. They'd probably got pretty good photographs when they did attack. That's what'll happen again. They'll have another go at us, I'm sure.' Despite such fears, by 25 November one could register a return of 'purposeful demeanour', an interest in getting the city back to life among many of those left or returning. Smiles and laughter were still noticeably sparse, but a thankfulness nearer to joy was quite strong. 'It's a miracle anybody's still living.' As diagnosed on the spot, however crudely, this sort of 'self-preservation' developed out of 'the apparent stoicism with which they have lost relatives and friends and homes'. A commercial traveller, who visited Coventry regularly, noted that week in his diary that most of those he met there were once more in good heart. After a period of feeling 'helpless at first . . . having got over the initial shock I think they are now prepared *to stand anything*', he wrote.

In terms of war production Coventry was crucial. Churchill himself, in a secret paper written shortly before Dunkirk, coupled it with Birmingham as 'vital' – which makes the attack's apparent unexpectedness all the more extraordinary. Some of the twenty-one important war factories hit on the Thursday night were put out of commission for a while, but never wholly. Most were able to carry on, on an improvised basis, yet more or less effectively. As A. J. P. Taylor has pointed out, the place was mostly back in full industrial production within five days. By comparison, in the corresponding period, the civic authorities took only a few steps towards recovery.[4]

This industrial recovery has to be closely tied to the profit incentive and to professional skills; and that depended, in this case, primarily on the willingness and ability of most workers and staff to get back to work. Despite all discomforts and physical displacements, they got back with speed. For family wage-earners simply had to earn. In those days of the early forties, Britain was far from a welfare state. Money, too, was much tighter, real poverty widespread. Fear of unemployment was a living reality for the labour force. Patriotism apart, the need to keep at the job was over-ridingly powerful. In the long view it can be seen as dominating all else, thereby destroying the basic premise in the planning of 'indiscriminate' bombing, that it should be aimed at the workers rather than the works.

The Germans, however, like the British and Americans thereafter, overestimated the lasting effects of each attack. Coventry was not hard hit again for nearly half a year (8 April). Yet Churchill thought the place could have been knocked out. 'They would have done much better,' he wrote (three pages after the passage that heads this chapter), 'to have stuck to one thing at a time.' This does not, naturally, presuppose that *anything* can knock out a people broadly satisfied with their own society. Is it that man can, under such circumstances, do as the commercial traveller's diary implies: stand anything? According to Basil Collier the German Air Staff got to know the British view of their own vulnerability: 'Yet they failed unaccountably to profit by their knowledge.' Nor was there any question of deterrence by loss. At the most two and perhaps only one out of the 449 bombers over Coventry was shot down, less than half of 1 %, a piffling figure beside 'acceptable' RAF experiences over Germany. Good weather at Coventry helped Fighter Command field 125 intercepting plane sorties. Of these only seven saw their enemy. Two opened fire, without success.

Remarkably, too, never again were the Luftwaffe to aim so

many planes at one lesser town. On London the record later grew to 712 (19 April), but outside that, Clydeside runs a poor third with 386 (5 May). Anything over 300 was unusual. Most towns felt mighty misery with far less, as in the November series 159, 121, 128, 123 for Southampton, the next town to suffer . . .[5]

7 The Southern Ports

(1) Southampton, 'the English gateway'[1]

'In early November the Luftwaffe made ready for a new stage of the air offensive. If Great Britain could not be bludgeoned into swift surrender, she might have to be worn down by repeated hammering. In any case everything possible must be done to check the expansion of her war production and prevent her from repairing recent losses. Air attacks at night would therefore be extended to the chief industrial centres throughout the country, and to the great commercial ports through which both everyday supplies and special consignments of war material reached her from abroad.'

BASIL COLLIER

The 503 tons of high explosive which fell on Coventry that Thursday night and Friday morning killed only just over one person per ton (about average). But the ripples of emotional blast ran through Britain. The onlooker role of the provinces was over. Diarists even in remote villages register refreshed if transitory concern. Still, a commonly expressed belief was that one's own particular community would not be so badly hit. Rich and varied were the reasons produced for this, as with our diarists far in the countryside near Wilmslow, Cheshire, who recorded on the day after the Coventry raid: 'our evacuee lady from Southend proposes to return home next Wednesday, as the war will soon be over! (15 November, 1940).' And return home she did, to the Thames estuary. Such were the complexities of human reaction, always to be recognized when generalizing today.

So to Southampton (population *c.* 180,000, or a little over 2% of Greater London's) on Sunday, 17 November, with 159 planes loosing off 198 tons of HE plus 464 canisters of incendiaries. 121 bombers came back on the following Saturday, 23 November.

Then again, with a week's lull, for two successive nights, Saturday, 30 November, and Sunday 1 December, 128 and 123 bombers dropped 1650 bombs and thousands of incendiaries, killing 214. Such concentration on a relatively small provincial conurbation inevitably produced profound social and other effects. As one observer reported three days after the second big raid, 'There was a fairly general feeling that Southampton was done for.' A middle-aged, middle-class man put it: 'We'll have to abandon the whole damn town.'[2]

Yet suddenly Southampton was left alone. The next serious attack, with only 62 bombers and relatively slight damage, did not come until well into the New Year, Sunday, 19 January 1941. Thereafter, only occasionally, though Plymouth, next port westward and first bombed in between Southampton's first and second big nights, continued to be hit hard into April (five furious nights) and up north-east Hull well into and past May (8 May, 120 bombers).

This erratic pattern in the German blitzes – with variants which we shall (where relevant) pursue – seems in retrospect to be nearly senseless. The sequences show no logic, no discernible theory of what such attacks – more or less indiscriminate bombing of all structures within a few limited areas nightly – were supposed to achieve; nor any reason why one place was left alone for weeks or months, while another was given serial assault, though still never with any consistency. This very uncertainty was, of course, one explanation of the lack of pattern. No one in Britain could know or predict where the next bombs might fall. Thus precautions and defences had to be kept diffuse. But the meagre evidence indicates that this uncertainty was also considerable in the minds of those responsible for target decisions at the Luftwaffe end.

This erraticism, which had a powerful influence on the continuing effectiveness of the raids, appears even more extraordinary when we realize that the Germans were themselves baffled by what seemed to them an equal incoherence in the allied air-attacks on Europe. Hitler's closest and most consistent minister-adviser, Speer, frequently refers to this phenomenon, regarding the Allied effort as thus rendered largely ineffective. As he saw it from Germany, the enemy had always demonstrated a lack of consistency, because they 'switched from target to target or attacked in the wrong places'. The consequent weakening of impact was enormous.[3]

The repetitive but erratic aspect is crucial to any understanding of what it was like to live through the provincial blitzes, raising fundamental issues of human adjustment which did not arise in the same way in London, a huge target with almost continuous bombing experience; or Coventry, at the other end of the scale, with virtually no continuity. As illustrated crudely by this table, Southampton was the first town to experience this type of cycle, which was later to hit Plymouth, longer and even more savagely.

BLITZ ATTACKS ON THREE PROVINCIAL CITIES, 1940–41

		No. of German bombers	*Tons of HE*	*Incendiary canisters*
COVENTRY	14 November	449	503	881
(1939 pop. *c.* 220,000)	8 April	237	315	710
		686	818	1591
SOUTHAMPTON	17 November	159	198	300
(1939 pop. *c.* 180,000)	23 November	121	150	464
	30 Nov–1 Dec	251	299	1184
	19 January	62	57	325
		593	704	2273
PLYMOUTH	27 November	107	110	170
(and DEVONPORT)	(13 January	50	21	749)
(1939 pop. *c.* 220,000)	20–21 March	293	346	1884
	21–23 April	354	403	2568
	28–29 April	286	369	1351
		1090	1249	6722

Southampton, the smallest of the ports to be heavily blitzed, was to end up Britain's ninth in total bomb tonnage received, one behind Coventry, with Plymouth fourth. Traumatic for the smaller southern port was the feature of that new post-concentration pattern, which may be termed 'episodic blitzing', including the first provincial 'double' (30 November and 1 December).

We were there from just after the first raid to long after the last, so that the record is strong around that gateway behind the Isle of Wight. The record is further usefully strengthened by five other main sources which round off the picture to an extent not available for anywhere else outside London. These are: (1) a local history with a strong blitz chapter (Bernard Knowles, 1951); (2) a 1973 thesis on this theme by a University of Hull geography student; (3) two collections of press reactions to disclosures of wartime studies by Mass-Observation and by the Inspector of Air Raid Precautions, Ministry of Home Security, both released in 1973; (4) the aforesaid Home Security report in the Public Records Office; (5) the continuing presence of a wartime M-O investigator, living in Southampton, regularly studying blitz memories as well as specific reactions under (3) and (4) above. These different sources are rich enough to make a book on their own. Here they can only be summarized more briefly.

It is easy to forget, after thirty-five years, that although there was no serious attack on Southampton until mid-November, the alert siren officially sounded 1605 times, while the town's first bomb fell on 19 June 1940. The last recorded all-clear went off at 8.08 p.m. on Sunday, 5 November 1944. 2631 bombs were recorded in 51 definable 'raids', which included 36 parachuted landmines and, later, one flying bomb. In all, 631 people were killed and over 40,000 properties damaged.

The earlier small attacks were mainly in daylight, as on 13 August 1940 when the huge cold store in the docks was hit, some 2000 tons of butter starting a blaze which defied 10,000,000 gallons of water.[4] Many smallish attacks followed, notably late in September on the Supermarine Airfactory, original home of the Spitfire fighter. The works were destroyed, though production had already been dispersed to 35 sub-units under the 'shadow factory' scheme. The local historian, Knowles, describes this as 'a blow which shook the town to its depths'. If so, it left little mark on people's minds after the all-out general assaults of November – which he describes as: 'indescribable.. even the most savage convulsion of Nature could convey no idea of the universal uproar and clamour. Every possible form of terror was used. Every second or two the town was shaken to its foundations.'

By the fourth mass attack, 1 December, 'it was as though the town had been the victim of a savage and brutal assassination', while 'everything bore the appearance of ruin and decay', and 'nothing remained that was not wilting, wasting or warped.' Thus Bernard Knowles, who goes on to eulogize the city's heroism under fire with equal extravagance. Thirty years later, Paul Nicholson, preparing his dissertation at the University of Hull, felt frustration as he tried to get a close-up human picture of what happened in his native city well before he was born there. Southampton sources emphasized always the best side, he thought. So he came to the M-O papers at the University of Sussex, which he presently found to 'form perhaps the most important contemporary material about the Second World War as they were not intended for publication and are free from wartime censorship', thus are 'very explicit in their accounts'.[5]

To look at what happened more explicitly, let us start with one report filed on 4 December 1940 after that 'savage and brutal assassination'. J. B. Priestley in his *English Journey* put Southampton in his third category, 'Twentieth Century' ('everything given away for cigarette coupons'). But the novelist and popular wartime broadcaster found it 'a real town, a town that has not fallen under the evil spell of our times', its people pleasing, 'well-fed, decently clothed, cheerful, almost gay'. Just before the war started, a great neo-classical Civic Centre had been completed at a cost of £750,000 – 'perhaps the most ambitious civic building created in the provinces in the inter-war years' according to Sir Nikolaus Pevsner.[6] In this shining white colossus, the civic servants were fully – vulnerably – centralized when destruction from the air hit at the place's cheerfulness and almost gaiety. But certain factors helped, at least at first, to soften the blows, as indicated at the start of our December report.

a. The damage to domestic and business dwellings is, so far as we can judge, more severe than at any other place yet studied.
b. The raids were violent for nights in succession.
c. But for some time before they had been occasionally severe, so that these violent raids did not take the civilian population completely unawares, and there had been much evacuation already.
d. The topographical factor is also important in interpreting the Southampton experience. Around the centre of the city is an un-

usually wide area of park, lawn and open space; there is not the densely clustered core of a city such as existed in Coventry, and was totally ruined there. In Southampton, the equivalent area is the main street, running through the Bargate. This has been shattered from end to end; but it is not in the same way the heart and soul of the town.

e. Finally, the population of Southampton is to a high extent genuinely resident and locally interested. Southampton has deeper social roots than Coventry or Stepney. There is a certain tradition of local toughness, partly associated with the docks and the sea.

Official accounts of the effects of the first blitz, even when not for publication, were already proving inadequate for an overall view, outside specialist departments. From December on, therefore, we included some general description of the physical landscape although this was not our direct concern. In Southampton, this was given with due apology – 'official sources often present the information in a somewhat complicated manner.' This is how it looked the morning after the first double attack:

The main shopping area of the town is practically gutted. Very few of the main stores of any sort are open. A large part of the business area is also seriously affected; only two of the main banks are functioning, for instance.

Most of the cinemas, many halls and churches, have also suffered severe damage.

Fire seems to have been a major factor in this.

The densely crowded small shopkeeper and working-class area east of the town centre has also been severely hit, and several acres of it virtually consist of debris.

The dock area has also been heavily hit, and a large part of the dock wharves and sheds are in a state of tangled chaos.

In all other parts of the town there is damage, and probably a majority of houses have suffered to some extent. But the working-class area around the ferries is not so badly hit.

The residential area on the west side of the town, especially in the Springhill district, has got off quite lightly, and is perhaps comparable to the position in Kilburn or Abbey Road, London, at the present time.

Factories have been severely hit, the Pirelli Cable Works is in a real mess; Thornycroft has suffered considerable damage, and one of the striking sights of the town is the huge Ranks Flour Mill – one side of the gigantic storage tower has been sheared away.

After such destruction we found the people themselves (*pace* Priestley) were 'not gay, they were not laughing; but they were not dumb or shattered or groaning'. There were few signs of anything like the scale of distress seen at Coventry, let alone hysteria. Many had left for the countryside however. Among those who stayed in town, only one real case of hysteria was observed – 'a middle-class woman who had a terrible scene with a Naval Commander in an hotel'. Tracking back to the original report of that episode, however, it hardly seems so 'terrible'. The relevant part of this document is, in its remote way, a simple statement of how far near-hysteria (as then seen) was from hysteria proper:

In the hotel ... residents had spent the two nights (30 November and 1 December) in a shelter together, tempers were frayed. At dinner one woman was talking in a loud voice of her fears, her sufferings, and so on; her son wanted to go out in the raid and park the car, and she was sure that he would never be able to get back. Towards the end of the meal, a Naval Commander, another guest, said to his wife in what was intended to be an undertone 'if that woman doesn't stop, I'll ...' (rest inaudible). Unfortunately, the woman heard, and a slanging match went on across the dining-room, the woman saying that she thought she was exceptionally brave, after all she had gone through; the Commander saying that it was too much to hear her all day and every day.

Later that evening, in the same hotel, a battery set was working (no electricity) with everybody anxious to catch the 9 o'clock news. As the announcer reached the crucial point – news of the Southampton raids – an elderly lady rattled her beads loudly, preventing another from hearing. 'She seemed a little senile.' After the news, the second lady came over to the rattler and said 'I'd like to murder you.' Those were the margins of Southampton post-blitz temper.

As in mid-November Coventry, so in ravaged Southampton at the end of the month, the commonest general comment went 'it's wicked' or 'terrible, isn't it?' This kind of verbal shock was not matched by any positive, aggressive reaction. There was little criticism, small grumbling, very small concern for the wider war and its future, and virtually no spontaneous demand for reprisals on Germany. Attempts to incite this belligerent response were pretty unsuccessful. Conversations with individuals and groups 'failed to provoke anything stronger than the wish to tie bombs round the necks of the actual Nazi airmen who had bombed Southampton'. As in Coventry, harsh experience on the spot made most people less rather than more ready to favour any equivalent indiscriminate reprisal.

This somewhat passive attitude did not necessarily derive from fear of further attack to follow if you shouted – as much of the press was doing – for revenge, since a large number of Sotonians felt that there would be no more raids upon them anyhow. 'There's nothing for them to hit now is there?' was a typical comment. This attitude helped those who stayed put. Many of these, too, showed a notably sustained interest in local bomb damage, particularly in a large crater near the Green where a lethal direct hit had destroyed a full public shelter. Talk overheard around the city centre was overwhelmingly on this subject. Similarly, the big department store, Jones, which had been gutted, continued for days to excite exclamations of 'shocking' or 'terrible'. The demolition of dangerous walls by the army attracted watchers, raised some laughs, and 'some isolated cheering'.

Sotonians frequently compared their experience with Coventry's. There was a strong if 'suppressed feeling of pride', of the competitive kind commonly noticed about London. 'On the Monday and early Tuesday people were saying that the damage was as bad as Coventry, but by late Tuesday this was developing into "worse than Coventry". One man said it was worse than Pompeii.' There was annoyance at national newspapers, felt to make insufficient fuss over the raids – they were heavily censored on this subject, of course. This pride in their own suffering, with emphasis on having it publicly recognized, was clearly a considerable help in keeping people a little more pleased with a superficially thankless situation. Under this stress they wanted not only to be able to feel (quietly) brave but to be recognized as such.

This wish-to-be-a-hero was a positive help to the anxious or half-shocked. After the worst was in fact over, the same process, modulating memory, produced in its afterglow a picture of total heroism, in Southampton and, indeed, nearly everywhere else blitzed. One local woman, now fifty, nicely 'remembered' the more masculine mood of her first visit, when she discussed her memories of 1940 in 1973. 'I remember (she says) coming here (first) on my honeymoon. My husband was stationed in Portsmouth and he found me a place in a little village outside the town. And do you know why he brought me into Southampton? Because he wanted to tell his mates whether the bombing was worse than in Portsmouth!'[7] On the other hand, the outlying villages, through Eastleigh and Chandlers Ford to Otterbourne, Twyford, Shawford, past Winchester and Romsey, fanned out across the New Forest, a radius of thirty miles and more, were flowing with unplanned evacuees, some bombed out of their homes, many more unable or unwilling to stay anyway. Here the air was more heavily charged with vocal aggression, including demands for 'violent reprisals'. This was to become a familiar phenomenon as the provincial blitz unrolled: 'go easy' from those sticking to the stricken centre, 'blast the bastards' from the physically unhurt periphery.

In Hampshire, this periphery was by early December, alive with concern. By the Monday night after the big raid of Sunday (1 December), a large portion of the population had left town for the time being, including nearly everyone from the inner core. Those who remained in damaged sectors 'began to feel pretty tough and brave', though with measurable undertones of anxiety – shown, for instance, in a high degree of gas-mask carrying, 32% of those outdoors in town.

The figures for those who did not stay put in Southampton could not be checked then and will never now be known. (Only 37% of eligible official evacuees had left in September 1939.) Published accounts usually ignore, and never assess, the role of unplanned, post-blitz evacuation, clear sign of an initial defeat among civilians. A survey showed 'whole streets deserted' by early December. By 10 December our estimate was down to 20% or less of the normal resident population sleeping in old parts of the city. A count gave only 3% of these as children, more than a third women (mainly young women). A great many had left homes only slightly (if at all) damaged. Others, mostly men, remained alone in partly-damaged homes, despite many difficulties; fortunately for them, the weather was not too severely wintry.

The situation was complicated by two different forms of evacuation: total and fluctuating. The 'totals' simply quit, flooding the country and scattering much further out where they had relatives. Some of these went north or west into Wales, and never returned. 'Fluctuating' covered a minority who mostly stayed away but sporadically revisited their homes to clean up, watch against looting and the like; and at first a much larger number who came back to town (to work) every day and went out again at night and weekends.

This second group were 'trekkers' in semi-official jargon. They clogged already disrupted transport systems. *But* they included the very many who kept at their jobs, returned to duty daily – commonly the man to factory or office, the woman back home – but moved off at dusk into the zone beyond the sirens' nightly wail. Dockers especially were making long journeys (e.g. from Salisbury) on trek. Taxis early on ceased to be available for casual calls. Many private cars, some taxis and a lot of buses became, in effect, mobile dormitories. Owners or customers spent nights sleeping on the seats and drove back in the morning. Bicycles were at a premium, as well.

No wonder, then, that an investigator leaving the train in Southampton docks on the evening of 2 December was 'impressed by the seeming deadness of the town . . . no cars and hardly any people'. Further out, there had been plenty of movement before dark. The country buses ran full to capacity; some had soldiers aboard them to prevent forceful overcrowding. Men and women were walking out of town, carrying all the baggage they could manage. Some were going to relations, some to outlying shelters, some even to sleep in the open. 'Anything so as not to spend another night in there.' Though many cars had space to spare, few responded to the numerous hitch-hike signs, thus causing much annoyance – as did the sight of coaches passing completely empty (presumably for dormitory duties elsewhere).

All this began and developed as part of a biological – an animal – response to events, a search for survival without any formal planning, let alone higher leadership or guidance. People, families or individuals, just decided for themselves, according to temperament and opportunity and chance. No clear official attitude could be detected in Southampton before

December. Even then it was, we found, far from adequate, falling far short of the massive social needs, at least until there occurred one of the first of several quite unexpected (and apparently otherwise unrecorded) effects of blitzing: a kind of *putsch*, with a determined junta momentarily in control. For in Southampton at this stage of stop-go confusion: 'A group of RAF officers virtually took over control of Central Hall, the main evacuation depot. They did good work and produced considerable order out of considerable chaos. (4 December, FR 516a).' The RAF so acted to express their solidarity with stricken civilians as well as to show that they 'were absolutely disgusted with the official handling of the evacuation'. Two investigators subsequently spent the evening with these officers and heard the whole story in detail. The airmen deeply criticized a breakdown (in their view) of this and other elementary systems in Southampton, despite its by now substantial experience of bomb attack. They were scathing about the leadership of the city as a whole. They considered that some of the top civil and civic servants had either panicked or failed by default to meet the needs of the crisis. Some, they said, refused even to see it as a crisis; these had quit intellectually, some of them actually, physically, too.

Those who stayed, regardless of risk (or in blissful confidence of better nights to come) did feel a sense of superiority over those who went. For a while, they deemed themselves the equals of any brave folk on earth. Why not, when their neighbours fled? Yet the Bishop of Winchester apparently found little to distinguish between one lot and the other. Dr Garbett, later Archbishop of Canterbury, came by car to Southampton on 2 December to find his people 'broken in spirit', with 'everyone who can do so leaving the town'. Everywhere he saw them with packages, suitcases, kitchenware, kids' toys, 'struggling to get *anywhere out of Southampton*'. He declared flatly: 'For the time, morale has collapsed. I went from parish to parish and everywhere there was fear.'[8]

It was not until thirty years later that the city's leading trekker was publicly identified – though it was privately a matter of harsh comment at the time, by no means only from the RAF: Public Record Office war files then taken off the secret list included a report mentioning, officially, that the highest political local, the Mayor (who died in 1965) left his key post around 3 p.m. each afternoon, to trek to his rural hideout. The next get-away train for him was not until after dark; he would not risk staying that late. At the same time, the Chief Constable – whose headquarters in the Civic Centre was twice hit on 23 November (but remained in action there until later) – was himself hurt the second night of the blitz, leaving an experienced but elderly deputy in charge. Knocked down by a vehicle in the early hours of the morning, he never returned to full duty at the soon regrouped police base in the Polygon Hotel. He retired the following year.

The third key man – and usually the most important – in any normal provincial set-up was (and is) the Town Clerk, the highest officer of the local authority, answerable directly to the Mayor (or, in larger groupings, Lord Mayor) and Council. This Town Clerk stayed at his post. But he was later described as moving about in a sort of coma. Dressed in a mackintosh, he went wandering 'from group to group in gloomy inactivity' when he was supposed to be making decisions over a large range of blitz and post-blitz problems.

By then it was 5 December, after the M-O report earlier cited. These personal descriptions are not just ours, but those of the highest national officer concerned, the Inspector General of Air Raid Precautions. In part because he was infuriated by our reports, the Minister of Home Security had sent his top man down to check – having left it primarily to the locals hitherto, apparently accepting their reassuring views. The IGARP was another man with an RAF background, that pioneer of precautions Wing-Commander E. J. Hodsoll, later knighted (1944) for his work in this post. He had the reputation of being an unflinchingly honest man, a martinet. His visit led to investigation and even intervention by the Regional Commissioner, who had powers to override local authority – though this high officer was also strongly criticized for failing to go (or to send his deputy) at once to help bolster up Southampton's wavering systems of controls.

The Regional Commissioner soon reported that the Mayor's Emergency Committee, now in charge, was indeed inadequate to the occasion. His Deputy Commissioner considered there was virtually no civic organization functioning effectively; various departments of the city authority had 'lost touch with each other', continuing uncoordinated in crisis.

Those working down nearer the roots did not at once recognize the degree of high-level collapse. But one saw it everywhere, in effect. Like tens of thousands of citizens inextricably, immediately involved, an observer could at first barely distinguish between dislocation due to the blitz or to subsequent incompetence in higher response. There was too little time and too much pressure to make such analysis profitable, even if possible, on the spot. Looking down, though, Inspector General Hodsoll saw clearly. He wrote after his 5 December visit, that: 'there was no outstanding personality in Southampton's local government at all.' We will return to these harsh words presently.[9]

No authority or organization could hope to offset all the immediate effects of massive air attack. Recovery could only be a matter of degree and time. No one could help much of what actually occurred on bomb impact. What mattered to nearly everyone individually as well as to the national effort as a whole, was, however, the scale and speed of action taken to overcome – or at least soften – the immediate effects. The slow pain of so many post-blitz situations has been largely obscured by a record which seldom shows the prolongation in time of the problems which followed the initial disaster – Southampton provided a fair example.

The greatest failure here was a failure in communication. This was not a matter of pretentious exhortation or so-called 'morale boosting' but, more simply, direct messages on matters of practical urgency. One example: Rest Centres, primarily intended for the temporary accommodation of the homeless, became of crucial importance in bridging the gap between total, domestic dislocation with possible personal panic and relocation in an orderly way. There were officially said at that time (November-December) to be at least 20 Rest Centres in Southampton. Investigators had difficulty in locating *any*. Typically, one had 'to ask nine (local) people before he got any direction to the nearest'. Local historian Knowles had '15 out of 21' working fully by 1 December. Whether they 'worked' or not, many had no pre-knowledge of them and failed to find – or even to think of finding – them when bombed out. Worse (for the authorities), other persons *not* bombed out but with some damage to their home or other domestic anxiety regarded such centres as equally for them; and no one controlled the inflow. Another investigator, after asking several policemen, located the main Central Hall emergency facility and found there: '. . . a great deal of argument was going on about whether people be allowed in or not, and periodically the doors were shut; despite the fact that police were *advising* people elsewhere in town to apply here for further particulars.' The flow of information was handicapped gravely, by the blowout of the main telephone switchboards (30 November); and by a temporary knock-out to the lively *Southern Evening Echo* (1 December). But the big post office by the Civic Centre, though not seriously damaged, remained shut, without posting any alternative address. Improvized information services became vital if the prevailing confusion was to be relieved, misunderstandings (and rumours) reduced.

On the Monday (2 December), about the worst time for much of the remaining population, a single van, with one forward-pointing loud-speaker, began to circulate after noon, repeating again and again that pipe-water was undrinkable unless boiled. This van circulated so fast that it was ineffective in getting across even this single message. Anyway, it soon broke down and was still not working on Wednesday morning when an investigator contrived to track it down. He found 'a number of policemen and civilians tinkering about with it'. Two other investigators present in town throughout did not hear any van, though they were on their feet all over the area through daylight and many hours of darkness.

Most water mains had been cut off anyway. Army water-carts were deployed to help out, apparently working on their own plan. But as gas and light were also off in most parts, *how* to boil your water as advised became another tiring problem. The fact that the authorities deployed one loudspeaker van compared with 75 fire engines from outside districts as far afield as Nottingham speaks for itself. The ratio 75:1 for material against social problems unwittingly gave somewhere near a correct outline of the priorities on that charred battlefield.[10]

Those who, bewildered, decided their home town was no longer habitable, were faced with a similar breakdown in communication as they sought to leave. It could prove difficult to escape the present mess. Transport services had patently been dislocated on a large scale by damage to vehicles, street debris and many diversions, including some for unexploded bombs. These last were not (we thought) cleared as quickly as in Coventry. But allowing for that, an 'objective appraisal of the situation strongly suggests that no imaginative or forceful use of the available transport facilities was made'. (On-the-spot reporting grammar.) People in distress were unable to find the starting points of buses or other facilities, already overloaded and diverted. Personal evacuation patterns became correspondingly chaotic. 'In this, and *all other* organization matters (except the water boiling), there was a complete lack of announcement, notices, direction boards or informed policemen to help the civilians.' The famed Civic Centre's central notice board continued into December devoting itself to adjurations on blackout behaviour, Defence of the Realm Regulations and a meeting of the Town Council! Not until Wednesday (4 December) did one new, post-blitz notice appear. It characterized the inadequacy of civic leadership in the face of these vast new experiences. The undated notice from the two top men concerned was weakly phrased, lacking in clear advice, and at two points liable to add to rather than ease confusion. It was printed ten miles away in Romsey, largely through the initiative of a Ministry of Information committee member and brought to Southampton by him on Wednesday, when we met him trying to find someone who would distribute it – he was relying, he said, on meeting 'a supervisor of ARP in Aircraft Factories' whom he thought would do. Eventually the police took on the job – and pinned a copy on the aforesaid board. Here it is:

PEOPLE OF SOUTHAMPTON

Although the town has been severely damaged it has not suffered any permanent injury. Some of the public services such as gas and water are temporarily interrupted, but the necessary repairs will be quickly carried out and the services will be in operation again in a few days' time.

The principal works and factories are continuing to operate, and they will be able to employ all or the majority of their ordinary staff. Everybody should get in touch with his employer or the nearest Labour Exchange in order to resume his work.

Temporary transport arrangements will be made in order that people may get to work without difficulty.

THE BATTLE OF BRITAIN MUST GO ON. ALL SOUTHAMPTON MUST CONTINUE TO PLAY ITS VITAL PART.

Harold B. Butler
Regional Commissioner
Southern Region
W. Lewis
Mayor of Southampton

The modern reader may imagine that to find a Labour Exchange, let alone one which was functioning, became a problem in itself. Many administrative buildings were out of commission, psychologically almost as much as physically. The Army Recruiting Office was manifestly gutted, deserted (no soldier to redirect one, no notice or advice). The Pensions Department looked habitable but was deserted. The Municipal Health Centre in East Park Terrace bore merely a scrawled paper: NO ENTRY DANGEROUS. It is difficult to exaggerate the sense of confused frustration one felt under such circumstances. Another angle is provided by four (out of thirteen) paragraphs from our account of 'Utility Services and their Social Effect' as they could be observed in Southampton:

Over and over again we found homeless, old, pregnant, ill, and anxious-to-evacuate people who did not know where to get the relevant forms or information. The comment of one woman is not typical, but not far from it:

'Everywhere you go they tell you you can't go there.'

The difficulty of finding out about things was increased when the military 'took over' much of the town; they, and many of the police drafted in, knew nothing about local conditions at all. There is a need for definite information officers as an essential part of A R P.

A minor point in Southampton was the amount of time and energy wasted in *walking about*; and over and over again coming to some road with a time bomb or some obstruction, having to turn back down a street and try another way. Much extra, though trivial friction was caused by this, which could have altogether been avoided by notices saying NO THOROUGHFARE.

When the local and much respected *Echo* bravely reappeared, with four pages, it contained no helpful local announcement; instead it featured a warning from the Home Office to industrial firms 'collecting information about air-raid damage', because this made them liable to legal proceedings!

By the following week-end the situation at the Civic Centre had improved. The main notice board carried details of dead and injured (with addresses). Some firms were putting up separate notices for their staff. There were no Ministry of Information or other placards, though three of the former were found posted outside empty shops which also carried Town Hall notices telling people who wanted to know about available recreation, evacuation or ARP to inquire at the Civic Centre, by now very busy at 10 a.m.

Conspicuous and (at the time) surprising was the seeming failure of church and other voluntary organizations to help fill this communications gap. One priest said it was useless to try 'because not many people in Southampton go to church ... (and) now all our premises have been bombed'. It was true that church and club-type organizations depended largely upon large buildings; and that these buildings, easy targets, often contained no resident personnel. They therefore became easy meat for even a single incendiary bomb. Surprisingly small damage to chapels and halls could cause them to be abandoned, whereas factories, offices and many homes went on functioning under far worse physical duress. The idea that God could not protect his special buildings played a part, in Southampton emphasized by the insistence on Sunday pre-dawn and post-dusk assaults.

A priest of the day declared that the raids seemed to him to have turned people away from the church rather than towards it. There was some evidence for this in uneasy admissions like these in December. An Anglican warden said: 'The bombing's put a stop to churchgoing. They were never very religious before the raids, but any inclination people had to go and pray then has now disappeared. I come in here (church) every day at lunchtime just for a few minutes private devotion but I've never seen anybody else (lately).' And a Catholic with a bombed church: 'The people here seem less to desire God now than before the bombing started. There are churches quite near that they could go to but they'd sooner stay away.' Prayer appeared to many inadequate, not helping while the raids were on or in solving the difficulties afterwards. It was not so simple as that, all the same. The whole business of man's inhumanity to man, epitomized in German raids on Britain, Britain's on Germany and presently America's on Japan, had a deep 'spiritual' impact, to which we will return.

In the midst of all this, the formal services, ARP, AFS, AA and other systems continued, though the first was heavily eroded by unplanned evacuation. In ARP, as in some of the above organizations and others like the WVS, important posts might be held by people with private means who could most easily afford to leave town or who lived outside in the first place and now found it hard to get in.

Opinion was unanimous in praising the Auxiliary Fire Services (AFS), but 'divided on the subject of AA defences' – though on the whole it was felt anti-aircraft could not have done any better than it did. The shelters were generally approved rather than otherwise. In Southampton, these were certainly unusually well-constructed. As noted in December 1940:

> Shelters in Southampton are more numerous and adequate than in Coventry, and definitely more strongly and carefully made. They have stood up extremely well. In many streets we found surface shelters standing in perfect order with dereliction around them and debris on top of them. In one case, the whole of one wall of a surface shelter had been blown out, but the roof had astonishingly remained supported by the other walls, and of the 30 persons in this shelter, no one was killed or very seriously injured, according to informants in the same street. There was general satisfaction with these shelters, and no demand for deep shelters was recorded specifically.

No coastal city visited before or after Southampton had appreciably better public shelter facilities. But they did not fully reassure the majority in crowded areas. On the next Saturday night, 7 December, six large shelters in town and at Portswood were visited. The outlying three were full, the rest from a third full to one which contained only a single Civil Defence worker. That week-end only five bombs fell in the whole region. (Indeed for the whole of December after the first, there was only one mortal casualty and one house destroyed, with similar figures for January.) Week-ends continued, however, to be the times of maximum anticipation and recourse to the shelter, as the result of Southampton's previous week-end experiences. One working man spoke for many more: 'These shelters are pretty good, but I don't use them as a rule. Only on Saturdays. That's the day they come over ... Of course they're not too comfortable, but you don't have to worry. I manage to get a night's sleep in them at any rate.' All investigators agreed that the most difficult *physical* part of the whole Southampton situation was food. 'Food,' we found, 'was the key-note.' The effort to acquire enough to eat absorbed a great part of public attention. It led to much minor frustration and built up dissatisfaction. If, as we wrote then, 'Coventry was something like a panic', then Southampton was something like a picnic: a nightmare one.

Bread was in shortest supply, but all rationed commodities were seriously short. Some shops were bombed out, more closed. Supplies to those remaining open became highly erratic. Very few cafés or restaurants were open. Communal feeding was, at first, highly unorganized. Gas, electricity, and water were all restricted. These difficulties were immense, as people realized. They did not so much complain as suffer. 'It's getting no hot food that wears you down,' said a docker, while a young woman, clerical, recorded: 'I feel *hungry*. I've only had two chocolate biscuits this morning. And tea. But I had to kid myself it was medicine before drinking it. It was made out of my hot water bottle.' Many went without anything hot for days.

By 4 December, St Michael's Church Hall was one of the few facilities in use as a canteen, though the expected aid unit had not yet arrived. This hall soon became quite important, simply because it showed signs of activity, even efficiency, when so many others were comatose. St Michael's provided one of the rare positive responses to crisis from *inside* the city. It turned out to derive from the initiative not of the Anglican parishioners but the Christian Science movement, which volunteered to instal and run a communal kitchen. The vicar agreed. As Christian Science's Lady Almoner later described it to Bernard Knowles, within the hour they were serving 'tea and sandwiches to famished people who had been in their shelters two-and-a-half days, afraid to come out'. By the 19 December it still looked to us 'a poor affair, though one of the only two centres in the town'. Nevertheless, it was there and it worked, albeit humbly. The good lady put it all down to 'the flood-tide of Divine Love', which 'carried us through all our difficulties'.

As hot meals became more available, prices ran around one (old) penny for a plate of soup or cup of tea, sixpence for meat and two veg. Meanwhile, the *Echo* was advertising best Downland turkeys, Christmas wines and spirits (with brandy at 7/– a bottle). One fur shop announced business as usual with 'a large stock of fur coats', suitable enough for folk shivering with more than cold.

Several mobile canteens (WVS, YMCA, Church Army, Australian) came in from outside early on. It was widely noted – and resented – that these tended to concentrate on cups of tea, with priority to uniformed personnel of all sorts. One mobile was taken into the courtyard of the Polygon Hotel, emergency Police Headquarters, and kept there for the police, though they appeared to have adequate facilities inside. None of the others worked on any detectable plan of priorities. They one and all sat immobile after arrival in town. Many did not find them.

Here, as with everywhere else in Southampton's post-blitz, the concept of helping people to help themselves was almost absent. Those powers-that-be had not yet appreciated that it was nearly useless to provide expensive facilities, static or mobile, for the distressed and dislocated unless these persons were told, loud and clear, where to find them. The facilities were inadequate in the first place; incorrectly used in the second.

In spite of this dislocation of habits, grumbling was manifestly low-key. Nor was there any loss in elementary pity. Cats wandering distrait among the ruins and appearing in the most unlikely spots caused much sympathetic comment from passers-by. There were suggestions that the cats, too, needed help; the council should bring round milk. Then they could also keep the rats down. Cats play a remarkably large part in blitz memories. After 33 years, a housewife, now 70, could only talk of pets whenever the war and raids were brought up. 'I remember the time I lost my cat. There wasn't any peace anywhere that night – and I couldn't find the cat the next morning. And then suddenly I saw her. There was a big old-fashioned mangle in our scullery and she had been crouching under that stand under the wringer – apparently too afraid to come out when I called. Amazing the instinct that animals have – to hide and lie low when there's danger.' Along with the dearth of cat and other food, and of hot drinks, colder ones ran short too. Some city pubs served beer by day; at first none of them remained open after dark. There was 'absolutely no entertainment or leisure pleasure to be had', day or night. By 2 December cigarettes were unavailable. Fag-begging became a feature of the local landscape.

Into this depressing situation came another outside intervention, a royal touch of pity came to the otherwise leaderless town. For security reasons unannounced, His Majesty King George VI arrived on 5 December. Outside the crumpled windows of the mammoth Civic Centre, he inspected a parade of Civil Defence services, police and selected members of the shattered but vigorous water, gas and electricity departments – 'a proud cross-section of the local civilian army which had defied the worst Hitler could do' (Knowles). The same pen reported the royal party as received 'with such fervent demonstrations of love, loyalty, and enthusiasm as in days gone by reserved for the occasion of an Elizabethan progress or the return to the throne of a King Charles'. As His Majesty passed through town, Knowles says, 'excited multitudes lined the wintry streets which re-echoed to volley after volley of cheers and repeated cries of "God save the King".'

Thus the post-war afterglow. It did not sound quite like that at the time – largely because the visit was poorly handled. Not even a loudspeaker van announced the impending arrival. Much of the route was unlined. The party passed almost unnoticed. When a larger public later learned of their visitors, usually from the radio, the response was far from uniformly enthusiastic. It was 'good of them to think of us' and 'I suppose they do a certain amount of good coming around but I wish they'd give me a new house', were among the more friendly comments. Others, mostly men, criticized such sightseeing, and said there 'wasn't time' for that sort of fuss ('too busy cleaning up and one thing and another'). Fairly typical: 'If they gave new furniture, good food and no fuss, we'd be truly grateful.'

December paled, and with it Southampton's attacks. The M-O

team went elsewhere, leaving investigators behind to watch human problems resolved, as well as to stand by for further assault. We rather expected that the continuing lull would bring a fast return towards normality, especially in the return of evacuees and a quick switch over from trekking to residency. But though only a minority now openly expected the Luftwaffe to bother with their town any more, a great many behaved as if, privately, they thought they might. In any case, most remained away. The unplanned evacuees and the trekkers were becoming used to that way of life. 'People have thoroughly adjusted their daily lives to the new circumstances – sleeping in outlying districts and coming into work or shop,' was the mid-December conclusion, on the new improvised Southampton pattern.

To aid these adjustments, the food situation was improving, though still with 'immense desire for hot meals' unsatisfied. Mobile canteens were 'still few in number', concentrating on tea and cake and staying only at fixed points. Criticism of troops for getting more than their share of such facilities was strong; and the uniformed men were being blamed for a persistent shortage of cigarettes as well as chocolate.

Behind an evident slowness to 'recover', at all levels from Civic Centre to trekker, lay the feeling that, although return of the bombers in force might be unlikely as things were, if things did return to near-normal, the Germans would know and come back in force to repeat the dose. This idea was to recur elsewhere. It also coloured some people's reactions to any sort of optimistic public statement about their place's courage, however much they might privately praise themselves for surviving such savagery. When later, in May, Churchill praised Plymouth for rising sublime above its suffering, plenty there were fearful that this praise would only incite the Luftwaffe to stiffen the dose. In Southampton by mid-December we reported the minority trend this way:

People constantly mention or imply that it is really useless to try to pretend that Southampton is carrying on properly again. The following comments from working men, who particularly feel this way, are probably typical of a very strong attitude underlying the Southampton situation. The same was found in Coventry. Little attempt has been made to deal with it thoroughly: 'I've never seen a place so much beat. There's not a thing working. The dockers look busy but they're not. It's all show. We couldn't carry on if we wanted to. But we don't want to. What's the good. They'll only come and start again.'

'You don't know what it is to carry on. People who have lived here all their lives have just discovered they don't know the way outside their own doorstep so how can you expect them to carry on?'

It took time for most to accept the comparative safety of December. Over and over, talk continued on the immediate past; they thought that they 'had suffered the most a human being can' (to quote an observer). They seemed to need some further stimulus, some sort of 'spiritual' urge to stave off the shambles.

On 9 December, Mass-Observation produced a report under the general title 'Aftermath of Town-Blitzes', summarizing experience so far. Using the current terminologies of our main client, the Home Intelligence division of the Ministry of Information, this was part of the summary for Southampton at this stage after more than two weeks without bombing:

1. 'Morale' in Southampton has distinctly deteriorated.
2. This seems to be largely because so little has been done to provide interest and rally local feeling within the town. But it is accentuated by the extent of the ruins and the paucity of human population.
3. The public utilities are still seriously affected, and unrepaired structural damage of a minor nature is still immense. Thousands of homes have broken windows and leaking roofs which make them extremely unpleasant, if not uninhabitable. Many more people have gone than a fortnight ago – e.g. out of seven houses whose larders were examined then, six are now deserted.[11]
4. The local food situation has much improved. The high price of candles, only form of lighting in many parts, seriously affects the poor.
5. Nine-tenths of all talk heard was *still* about the damage and the raids of a fortnight ago. This topic remains an obsession, and among many people is becoming dangerously near neurosis. There are apparently *no* official attempts to provide any antidote or to make any attempt to extrovert these feelings.
6. The only touch of gaiety found in Southampton was a pub which had a pianist and a singer. This pub was congested and did a roaring trade and had a higher degree of cheerful conversation than any other pub – most of the others were practically empty.
7. The alleged dis-organization of the authorities and the failure of the local authorities to keep their heads has now become a subject of

comment in the surrounding countryside. The notice urging THE PEOPLE OF SOUTHAMPTON to return to work was displayed all over the countryside as well as in the town. Removed from its Southampton context, it created a considerably different and even alarming impression. [This notice is quoted earlier above.]

8. No major grumbles seemed to dominate Southampton, and, as previously noted, the shelters are generally regarded as reasonably satisfactory.

Complaining (as at no. 8) was rarely part of the blitz aftermath. Complaisance was a form of 'shock'. But looking beneath the acceptance and that phlegmatic surface we had to reach a rather harsher conclusion – maybe a little too harsh, seen in hindsight; maybe not.

9. The strongest feeling in Southampton today is the feeling that Southampton is finished. Many will not say this openly, but it is a deep-seated feeling that has grown in the past fortnight. Yet many householders continue to come in every day, and quite a number of women spend the day in their homes and the night in outlying billets. No special bus service has been organized to transport workers to and from outlying areas, even as far as Salisbury (72 miles). The 'instinct' for home and local associations remains, and the feeling of despair about Southampton could surely be much reduced by local leadership, propaganda, and some brighter touches.

In any case, the feeling that Southampton was, for the time being, finished did not in itself mean that Sotonians – let alone Britons – were finished. Maybe they could no longer be so fully Sotonian. But they could and did carry on being people, *homo sapiens*, albeit in varying degrees displaced.

1941 came in with no further serious local assault. London continued as an intermittent major target – 5000 of her 19,000 tons of H E fell after 14 November. Southampton suffered at the level of what could then be rated minor incidents – though, in peacetime, any one lethal bomb would be rated major tragedy.

Trekking, only a temporary device in Coventry, had settled into a way of life for thousands here; it continued as a strictly private, unrecognized, unled daily mass-migration. Inside the social structure of Southampton – and Britain broadly – it was not difficult, from above, to overlook the distress of the less fortunate; for that had been the practice of many years, to an extent inconceivable in these days of welfare state. Even the socialist elements in the wartime 'National Government' found it difficult to bridge this gap intellectually, to realize the full scale of ordinary unhappiness. Some of the misery and misunderstanding could certainly have been alleviated by more rapid, more drastic recognition of the full range of minor human problems.

Consider this passage from that most socially thoughtful of war historians, Richard Titmuss: 'The social problems arising from air attack in the provinces, in Scotland and Northern Ireland, differed in no fundamental way from those which faced the authorities in London. In one important respect only did they present a singular issue of their own, and that was the phenomenon of "trekking" which emerged to worry the Government in the spring of 1941.' But the 'spring of 1941' was a good half year *after* the phenomenon had 'emerged' to worry whole populations around the islands – for it was soon (and more dramatically) to be seen in Portsmouth, Plymouth, Swansea, Greenock, and many other towns. Trekking became, in a series of separate situations, one natural response to bombing; for many, it was the only response that they could devise for themselves, where no one else could be seen to be offering any clear alternative. Yet the Civil Defence Committee of the War Cabinet, specifically naming Southampton (and later Plymouth) as examples, concluded that this sort of thing showed weakness and therefore should be, in effect, regarded officially as a huge non-event. To quote Dr Titmuss again: 'The Government, interpreting trekking on this scale as a symptom of lowered morale, was anxious that nothing should be done to encourage such movements. No specific provision was therefore to be made for the people taking part.'

It was decided, at the highest level, that 'no official arrangements are to be made for persons who are not homeless'. What a contrast to the preparation of many thousands of beds for the shocked and bomb-sick! But then, that had been predicted and planned for; the electorate merely failed to respond. Instead of openly 'lowering their morale' by going crazy, they went privately to what was felt as fuller safety to sleep by night. This tendency was not anticipated and therefore not acceptable.

Such a bland approach to 'responsibility' for citizen welfare is hard to visualize now. It is there, writ large not only in the official histories, and in the Public Record Office, but at the other end of the microscope in these our field reports. It is almost as if the authorities, having vaccinated and inoculated their population against smallpox and cholera epidemics, instead unaccountably got a pandemic of bubonic plague – but refused to treat it, as being outside their terms of reference and, anyway, too lowering to contemplate.

The non-homeless, however defined in Whitehall, were expected to endure without fuel, light, hot food, drinkable water, cigarettes, entertainment and the rest. More important than that, they were supposed to stick it out without precise information or adequate guidance. Any intelligent person after a time could see that this and a good deal of the continuing mess was not simply the unavoidable result of blitzing, but showed the failure of imaginative leadership – whether local or national – to provide the essential responses which would re-establish the present and rekindle hope for a better future. Today we know what was then unknown: that a significant part of the local failure to respond was sanctioned, if not actually ordered, by the Cabinet down. And, if the minister in London implicitly disapproved of the Mayor's nightly trek, it was not publicly condemned. Everyone could please themselves (in the absence of the faintest expression of the official attitude) where they went and how – in contradiction to the basic philosophy of disciplined control in total war, which more than ever before told 'everyone' what they should buy, eat, believe and say, where they should work, or when they should be called to take up arms.

By the same token, the Southampton fire service had, not long before the blitz, changed its hydrant fittings to a more modern device. When fire engines poured in from all over the south to help out, many could not use their hoses because their couplings would not grip on the local ones. This was to happen elsewhere too. But it was not until May 1941 that a National Fire Service was set up to replace the mass of unco-ordinated (though individually efficient) local brigades. Here, as often, the problem was not exactly ignored; its solution was delayed, with tremendously damaging effect, largely because of the traditional respect for local autonomy. Much of what did not happen in Southampton stemmed from other aspects of that same respect, leaving it to the city to sort out its 'own' problems as the individual trekker had to do.

Trekking can be seen as a failure of personal or familiar will under stress. But to trek was not cowardly, defeatist or disloyal. It was a sound response to environmental collapse, an ecological re-adaptation to disturbances on an unprecedented scale. To get that effect, the number of homeless, the number of houses even slightly damaged, is not necessarily of first importance. Indeed, many humans can abandon their habitat quite easily, even in the safety of peace. An opinion poll across Britain indicated that, as of July 1974, some 4,000,000 Britons are eager not just to move lodgings but to move country – to emigrate. A still larger number are considering the possibility. New analyses of the 1971 Census of the United Kingdom show that, through the sixties, there was a major movement from cities into the countryside. London had a 7% loss in a decade, Glasgow no less than 16%, in marked reversal of all pre-war trends.[12]

In so far as, clearly, government wished people to stay in habitable dwellings, trekking was a defect in leadership all the way down the line. It demonstrated an inability to understand the needs of ordinary people in extraordinary times. Moreover, by belatedly recognizing trekking in May 1941 (when the needs were nearly satisfied), the truth was admitted. The policy had been wrong, in the highest judgement. Looked at from below, it seemed to be, to say the least, heartless. What then really was the point of the king or the prime minister coming to look at the destruction afterwards? In a sense they were trekking, too. Imagine the effect if either one of them had decided to demonstrate official policy by staying the night in an average-sized home – or brick surface shelter – in the city's centre; or in the Civic Centre for that matter. Yet nothing less would have made the government's point: don't move, unless you must.

The last bombs of Southampton's winter came down on a Sunday, 19 January, scattered very widely from Woolston and Bitterne to Portswood and St Nicholas: seven houses destroyed, a few over a hundred damaged, with one death and three more severe casualties. In all of February, nobody and no building was to be hurt.

By the latter half of January, many more signs of new life were burgeoning, especially through private enterprise. The Southampton Chamber of Commerce had even produced a 24-page Emergency Business Directory (price 2d). But a great deal remained improvised. The Fairway Corner Fire Party, with its headquarters at 157 Bitterne Road, required all local occupiers of property to contribute 10/– to a 'fighting fund' and thereafter 1d a week. Their circular letter of 21 February set the atmosphere before the start of spring:

21.2.41

Dear Sir or Madam,
The Committee would welcome volunteers to do the following work to the headquarters:

(1) Tidy up the garden.
(2) Paint the lower half of the outside of the premises.
(3) Clean and decorate the passage way.
(4) Redecorate the sleeping room.
(5) Clean and redecorate the kitchen.
(6) Cut access holes in fencing.

Also the Committee would like to know whether the ladies would co-operate in keeping the headquarters tidy for an hour or so a day. The following articles are needed, and the Committee would be very grateful to receive any as gifts or on loan.

Coal	Saucepan
Poker	Cups & saucers
Wireless	Cutlery
Floor rugs, carpets &c.	Rugs
Kettle	Clock
Black-out curtains	Tea cloths
E.L. Bulbs	Mirror
Pillows, cushions	Sweeping brush
Glasses	Dartboard
Towels	Pack of cards
Enamel bowl	Magazines
Milk jug	Table
Teapot	Torches
Sugar basin	Brandy
Frying pan	First-aid outfit

Yours faithfully,
Jasper Beazley
Chairman.[13]

Most public facilities were working more or less satisfactorily by the start of March, after weeks of what a resident observer termed 'comparative quiet'. Transport and telephone systems were still far from adequate, though, and the subject of many complaints. Typically, outside the Civic Centre Post Office the telephone boxes bore a notice FULL SERVICE IS NOW AVAILABLE FROM THIS KIOSK. Only three worked, and you had to dial O for Operator to get anything. As someone had scribbled inside a box: 'WHAT A PHONE SERVICE!'

Soon after dark, there was no public transport. 'Streamline' taxis would go out later, but were not bookable by telephone (if it worked) after 6 p.m. Night-life was at correspondingly low ebb. Cinemas had to close by 8 p.m. The Grand Theatre remained wholly closed. Pubs in working-class districts did fair business, with conspicuously few women around. But here, as all over, the main activity was by day. As the barmaid of The Winning Post, Bitterne, remarked: 'We don't open the lounge bar, except at week-ends now. Those people who haven't been bombed out have moved, and the people who're left go away at nights usually. It isn't worth putting on 16 lights and lighting a fire, just for 4 or 5. Most of our customers have evacuated.' As the lady makes clear, more than three months after the last significant attack, trekking and evacuation remained to dominate the scene. Towards the city centre that scene was distinctly parched. The open pubs, as before, seemed lively oases in the centre of town. By day, several central pubs were full of labourers (from outside), still clearing the November debris: 'The piano is often playing and an atmosphere of cheerfulness and noise ... *in the midst of desolation and deadness.*'

Dancing was the only other leisure-pleasure to prosper. Banister Hall, opening every week-night, 7–10 p.m. (admission 1/-) was installed in a private house near the Sportsdrome, whose clientele it took over. By 8.15 p.m. on 1 March, another Saturday, this dive was filled to overflowing (*c.* 3000), with about half the men in uniform (many drunk):

It was difficult to get inside the door for people jammed in the tiny hall, streaming up and down the stairs. Three large rooms were in use upstairs, one for taking tickets, cloakroom, etc. and adjoining were soldiers, sailors, etc., and girls sitting at tables. The dance hall itself consisted of a large room, with band on small platform, crowds of people round the walls, and about 100 dancing. Girls were young, C-D class, many in jumpers and skirts, sometimes flowers in their hair, heavily made-up, or in afternoon dresses.

The floor was so crowded that most couples just shuffled and hugged (in those pre-twist days). A move to the more ostentatious Guildhall seemed appropriate for comparison. This required a half-mile walk in the blackout, searchlights fingering the sky, a plane in the distance. About 400 people were inside (paying 2/- each). Before a count and age analysis could be made, there came an announcement from the bandstand, asking everyone to leave 'quietly in small parties' as a raid was expected.

In the Ladies' Cloakroom a crowd of girls besieged the counters, 7 or 8 deep, fighting to reach the front. As each girl obtained her coat, she had an even worse passage to fight her way out again, and the crush was intense and stifling. Overheard were indignant complaints about the necessity of leaving, the disorganized rush, etc.:

'They ought to give us at least 6d. back, I don't suppose they'll ever go on with it later.'

'I think it's an unnecessary precaution. There was nothing happening outside...'

'Isn't it awful here? They ought to arrange things better. What if we wanted to get out suddenly?'

After about 10 mins, the fight subsided a little and girls were leaving. Upstairs in the vestibule crowds of men were standing in darkness, while a steady stream were going out. People were standing about uncertainly, there was no one to be seen in any official capacity, no instructions or advice given.

The thwarted dancers proceeded to walk home for up to three miles. As they went, they talked of the Saturday night routine; of the eerie effect of seeing the night sky through the silhouettes of ruined churches. At the first bridge, and again when passing a railway station and a gasometer, they ran. Expectation of attack and danger were discussed fully though 'in a bantering manner'. None tried to hide nervousness at sudden flashes, Very lights and cars passing with headlights (improperly ablaze). 'Put that light down', 'lights' and the like in angry tones. Several times, approaching car noises were confused with planes. These cars had plenty of vacant space; but none would stop for the hitch-hike sign.

Gunfire in the distance, and about 9 p.m. bomb crunch. Saturday, which had started with a noon alert earlier, lived up to its reputation. False alarms had become the order of the Southampton day, as an observer's log illustrates:

Date	*'Alert' time*	*Activity*
20.2.41	12.15–1.0 p.m.	None
	7.40–8.40 p.m.	In cinema – no one went out
	9.15–12 a.m.	Little gunfire No incidents
21.2	11.00–12 p.m.	None
22.2.	1.45–3.0 p.m.	None
25.2.	12.00–12.30 p.m.	None
28.2.	2.15–3.0 p.m.	None
	8.15–9.0 p.m.	None
1.3.	12.00–12.45 p.m.	Gunfire, raider chased
	7.30–10.30 p.m.	Gunfire, bombs (2 killed)
	10.50–11.10 p.m.	None

Most alerts passed without incident. Up to Saturday, 1 March, for weeks no one had been seen to enter a public shelter during a day warning. In shops, post offices, cafes and streets, people had gone on regardless, though occasionally 'women would walk quicker'. At the Saturday noon warning, soon followed by a faint air-trail and puffs of smoke high overhead, clusters of uplookers formed upon the pavements. Many came outdoors to peer. At the first sound of gunfire, there was a measurable stir, until:

A number of women who had been near shelters now stood at their entrances. An old woman started to climb down the steep steps of a shelter under shop buildings at the corner of the New Rd. and four other women, two elderly, two younger with children, stood at the door, talking.

More gunfire was heard, and all the women went inside the shelter and sat down in a dark corner with their children, chatting together cheerfully, and exchanging bomb stories, etc.

Outside, one woman, *c.* 45, went on talking to the warden: 'No, I'm not going down. But after having my husband blown out of bed ... when I hear the sirens now I'm all like this ... (hands shaking). But we've got to keep ourselves in hand, haven't we? The doctor said to me afterwards, it's no use you being like this. But it makes you nervous, after you've been bombed once and blown out of bed.' Other shelters held people after a long spell of emptiness. Tunnel-type park shelters were mostly empty, though women, children and prams stood near the entrances. No rush, only quiet caution. Well before the all-clear, with no more aerial sight or sound, 'the life of the town was going on as before'. But that afternoon, the number of women carrying their gas masks suddenly doubled, from 26% that morning; next morning it reached 68%, plus 40% for men, high figures – a sure index of underlying anxiety. Through February the local *Echo* had been urging the need to carry masks at all times – with no visible effect, although a minority never moved without them. One woman carried anxiety so far that she was overheard at a bus stop declaiming: 'I haven't undressed properly this three months.' 'Nor have we,' confirmed another.

Two girls overheard on a bus echoed the views of those who had moved out and had no present intention of returning to town:

'We've got this house out here, I feel much safer out of Southampton, don't you?'

'Oh yes, I don't think I could go through that again, do you?'

'No, I don't think I could. Though when you think about it afterwards, you think, perhaps it wasn't so bad after all.'

'Oh, but it was terrible, wasn't it? I hope we don't have to go through it again. I'd never go back to town, would you? I'd much rather be be out here . . .'

And a middle-aged woman spoke of a friend who was shocked by the sight of her native town, and wanted to go back to Bath: 'She said she could have cried when she saw all the shops down and everything, she didn't know where she was. She wants to go back to Bath now, although Southampton's her home town. Oh, I wish the whole thing was over, and we could go back to things as they were, if only there were no bombs and raids it wouldn't be so bad . . .' While a middle-aged housewife staying on near the city outskirts in a small detached house, patched up after much local damage, summed up for the anti-trekkers: 'I just sat down and cried. You can see what it was. All the windows out, the front door and back door right off, glass everywhere, my lovely plant with all the blooms torn off (laughs). It was a terrible experience. I've never had such an experience, I don't think we'll ever forget it.' She described her experiences the first time the raiders came over, when she was alone in the house, continuing:

But there's one thing I always think, we've had the worst, it *can't be* any worse. Of course some people think we shall have it again, and some don't, but I don't know. I don't see what use it would be. I think, well, we've been lucky, all our neighbours are gone, this is *the only house occupied now down this side*. I don't sleep here, I sleep over the road, but my husband does. I think if it's going to come, it will come, wherever you are. I read in the *Daily Mirror* this week of some people who spent about £50 going all over the place, and they got bombed out three times. Each place they went to they got it. And then they went back home to find their house still there. That's why I don't want to move. My husband doesn't want to, he's got his job here, and I don't feel like moving in case we got it just the same somewhere else.

Through March small businesses and private facilities for public use slowly picked up. Commercial restaurants were running at low key. Most only did lunches. The Tolia with curried turkey at 1s. 6d. and the Michel with a 4-course lunch at 2s. 6d. (soup, macaroni, veal and bacon, sweet) prospered, as did the Tudor offering plain lunch at 1s. 5d. or 1s. 10d. with soup. At the 'smarter' end, the Dolphin Hotel, with 3/- lunch (or duckling and fruit pie at 2s. 6d.) was nearly empty, as also Gatti's, where they only had stewed steak at 1s. 6d., no sweets.

Although Sotonians no longer commented constantly on bomb damage among themselves, they unfailingly delighted to spell it out to strangers – what used to be where, when that was hit, particular incidents, victims: 'and that used to be the Public Library . . . and that was the fur shop, and Boots and – I forget what that was'. A growing sense of irritation came with the slow rate of clearing the ruins. As a middle-class housewife declared: 'I hate Southampton. I hope they *do* something' – to fix it up.

The local press expressed concern about the rates, risen sharply. Many businesses had been reluctant or unable to reopen. Many customers had moved out. Supplies went on being short, over and above necessary wartime restrictions. And a notice on the door of a chemist's shop read:[14]

WE REGRET WE ARE UNABLE TO SUPPLY

Vacuum flasks
Lipsticks
All tubes of vanishing cream
All barley sugar sweets
Rolls razor blades
Gillette razor blades
7 o'clock razor blades
Saccharines
Rouges
Rolls razors
Brushless shaving cream
Nivea cream

Goods in stock

Gardenia vanishing cream
Gardenia cold cream
Jasmin face powder
Jonteel

'My word, what *have* they got?' a young man asked outside. The March 1941 report saw this side of the Southampton situation as gloomy economically. The week-end inrush of shoppers from all around, normally a major source of business, no longer operated. At week-ends 'the town is almost empty', as the Saturday-fear of weekend attack persisted. A number of people now stayed the week but went away every weekend – to relatives, friends, lodgings, anything to get out for a spell.

Not that any of these categories are complete or conclusive. For instance, strikingly, we found some people who had moved *to* Southampton in March, because they felt so sure it was now immune from further attack. One young man insisted:

'I never hope to have to go through that again. I'm not afraid to admit I was scared, really scared. I thought my last hour had come. But Southampton's the safest place to live in now. Why do you think people come from Bournemouth and so on to live here? They're afraid they're going to get it next, and think Southampton's had its blitz. There's nothing to come for now, you can see – their bombs would just fall on places that had already been bombed flat. If you think about it . . . they sent over 250 planes, didn't they, and dropped 600 tons of bombs? Well, they're not going to waste all that again.'

He and his Bournemouth friends were not quite right, though not far wrong. Southampton, after another raid on 11 March (70 bombs) was one of the safer places in the south for most of the rest of the war. The city ended up with this record:

SOUTHAMPTON RAID EXPERIENCE IN WORLD WAR II[15]

Year	*High Explosive (On target) Bombs*	*Tons*	*Incendiary (nos.)*	*Casualties Dead*	*Severely injured*	*Property damage Total loss*	*Very bad*
1940	2117	280	13,802	481	672	575	1445
1941	388	153	12,400	103	151	269	1004
1942	111	37	4,450	42	60	68	111
1943	15	5	0	4	15	2	26
TOTAL	2631	475	30,652	630	898	914	2586

But 271 of those 481 dead of 1940 perished in a fortnight from mid-November, while 1651 of the major property losses fell in that briefer period.

The total in death and grave damage was patently small in relation to the total dislocation and distress. These lesser evils cumulatively added up to far the greater part of the blitzed equation. Less than 500 tons of explosive spread over the years may not seem a very heavy load for such a city to bear, remembering that in one night at Dresden 135,000 Germans were to die from Allied bombing and in Greater London nearly 30,000 were killed from the air. (The ratio of Sotonians killed to wounded was 1:3, near the national average.)

What hurt Southampton was its relatively small size and the concentration of major attacks within a few days. The pain was then prolonged by some smaller attacks and many overflights by aircraft otherwise engaged. What might have resulted from further large attacks in, say, late December and the New Year? We shall be better able to answer that question after looking at other places which had such experience.

But below and beyond such local factors lay the central theory that most of these *sorts* of civilian problems should, indeed must, be left not only to action by the local authority but also to local authority thinking before and decision-making after each event.

Southampton proved to be 'typical' of the local authority's inability to cope. Despite major theoretical preparation, the city was, in the psychological sense, unprepared for what happened in November 1940. The actual, physical effects of those nights was in important ways less than expected. The less tangible, psychological effects were far greater. The lessons of London had demonstrated this – in London. The local authority, with proud and essentially *local* traditions, local outlook, failed to benefit by this experience nearly as much as might have been expected by anyone looking at the social facts rather than the political and administrative. As Titmuss has so well shown, *initiative* was not a distinguished feature anywhere, even inside London. As he says, too, the surprising ('perhaps even astonishing') thing to those on the inside and 'most intimately concerned' was that post-raid relief services worked as well as they did, since: 'They were conceived without much thought and less money, they were nearly suffocated by the uniform of the poor law ... they had been neglected by Parliament and by the press of *all* political parties.'[16] As the official historian himself points out, the services needed, in 1940, to be 'informed with a new spirit'. During and after what he calls 'the stimulating experiences of 1940–41', the whole approach did change, but slowly. Instructions to the Assistance Board, for instance, 'began to be coloured with significant references to the need for courteous and sympathetic behaviour'. They had not been there before.

What remains surprising is that these inadequacies were permitted to persist into 1941, and beyond. This was the outcome largely of the way Sotonians switched their pride from being members of a peaceful community to a ravaged one. That intense interest in destruction, coupled with a conviction no one could have suffered more and, by inference, endured better, enabled nearly everybody to accept things as they were. This made it easier for stunned leaders at Civic Centre level to stagger on without radical rethinking, to remain narrowly 'conservative'.

If it had been practicable to put new leadership into Southampton on, say, 1 December 1940, possibly things could have been adjusted much more quickly. Inspector-General Hodsoll's report a few days later, as already noted, actually criticized the Regional Commissioner, as higher officer responsible for overall decision in this sector, because (Hodsoll considered) he should have gone there at once and helped bolster up the local authority. His Minister, Herbert Morrison, a socialist deeply experienced in local politics, was himself evidently reluctant to overrule any elected mayor or council, duly appointed Town Clerk or Chief Constable.

The cracks were, therefore, either ignored or papered over. The papering was made enormously easier, of course, by arguments of national security. Local journalists could not (and did not try to) tell it like it was. Public opinion could exercise no clear pressures. Reports of what was happening as seen by impartial eyes, Hodsoll or Harrisson or whoever, were for higher minds only, if read at all. They have only come into the daylight with the third of a century lapsed.

Not that Southampton, for one, has anything to be particularly ashamed about. As Mr H. V. Willink, Special Commissioner for London all through the blitz and able politician in his own right, pointed out in the House of Commons on 11 June 1941: '*Nobody* in my office, or myself, has *ever* been asked for any general information on the way London attempted to deal with the homeless problem by any local authority in England, Scotland or Ireland.' The lessons of London, painfully learned then, had to be re-learned (or not) in every other city. Let us look, next, at how some others on the south coast fared, each in their separate way ... in each place emphases were different, sometimes strikingly so thanks to local traditions, politics, geography, size, cohesion, and a lot of chance in where the bombs fell and on what.[17]

(2) Portsmouth (the Nelson touch)

'It's my belief the Government's brought this on us. And then they don't help you. They all ought to be hung. We'd be better off.' (Housewife)

Two artisans greeted each other in the street:
'Good morning.'
'Good morning. How are you?'
'Oh, I'm all right.'
'– Mostly down I suppose?' (laughing)

A conversation between the landlady of a pub and two customers, man and wife. The man told how, on Monday evening, he was coming for a drink and got shrapnel in his leg. Landlady and barmaid gathered round, interested, then:

Landlady: 'You're properly in the war.'
Woman: 'The war's right (turning to husband); you might just as well join up and be done with it.'

She spontaneously added: 'Isn't it awful? I'm fed up with it all.'[1] Those were fragments from mid-March 1941 in Portsmouth, the south coast naval dockyard town a few minutes flying east up the Channel from Southampton. It had a 1939 population of a little over a quarter of a million, nearly half as large again as Southampton, with which it had then few civic or organizational links except that both lay in the County of Hampshire. Portsmouth was an autonomous city.

We left the Southampton scene in mid-March, when to all intents and purposes its heavy blitzing was months past. Portsmouth's began when its neighbour was effectively cooling off, on 5 December (74 planes). There followed three major attacks, widely spaced – 10 January, 10 March and 17 April. These last two both involved higher concentrations of German bombers than in any one night on Southampton: 238 and 249 respectively. Portsmouth was one of the few provincial towns anywhere in Britain, which, once blitzed, was never doubled – Coventry was the most conspicuous other example, with an even longer gap (November to April). The intensity of those single raids, however, was enough to put Portsmouth above Southampton by one place, as number 8 on the major assault chart, outranked in the south by Bristol and Plymouth.

It so happened that the physical damage in Portsmouth early on was concentrated around conspicuous, socially regarded landmarks at the centre. In smaller towns of tradition, overnight obliteration of lifetime symbols of solidarity comes as a special shock. The visible, concentrated loss of familiar buildings can have far more impact than greater damage more widely diffused, as in London and presently the other great conurbations. The 'heart' of a small place was blown up and out, that way. In the mid-January Portsmouth attack: 'The almost complete destruction of one of the main shopping centres, and the striking spectacle of the gutted Civic Centre, continued after a week to attract observation and detailed comment. Whenever an air-raid smashes into the heart of a city, the effect appears to be far greater than where the damage is on an industrial or residential district.'

Early on Portsmouth emerged (on this record) as psychologically more upset than the attack seemed to merit by comparison with other places. By January 1941 the 'people have been seriously shaken', it seemed; so that 'further raids might have a far-reaching effect if things are allowed to continue as at present'. 'As at present' meant, primarily, the 'complacent attitude on the part of the regional and local authorities'. This complacent attitude was seen to be based, in turn, on 'inadequate appreciation of the problems which the *poorer* sections of the community (then) faced', and which hardly bothered the better-off (who were also leaders and administrators). There seemed to be a blind eye to the telescope in the city which still shows Nelson's flagship, the *Victory*.

For one thing, those better-off were considered by some to have *begun* the trekking to and from Portsmouth, where it became particularly conspicuous because of the narrow isthmus connecting the main part of the port-town to the southern mainland. The economically favoured were said to be driving out nightly after the first major raid, regularly refusing available space to others on foot. As the buses were already overcrowded, this infuriated the less fortunate. Counts of private cars confirmed this. Past a 50-yard long bus queue before dusk, 38% of the outgoing cars had three or occasionally more passengers, 55% had two and 7% only one. All were heading for the wide, well-populated back country of south-east Hampshire and adjacent Sussex (the Fareham-Chichester area).

So, by the start of 1941, Portsmouth had followed the Southampton cycle of trek and evacuation. The surrounding villages were crowded, with consequent 'extreme strain' on transport, food, health services and rural goodwill. Over a half-circle 10–15 miles along the coast east and west (shorter towards Southampton's overlap and up to 25 inland) city-dwellers were clogging every facility. Back in town, some undamaged streets were practically empty, while in others practically everyone stayed until later in January. A study begun then and continued over the following six months, covering four selected, poorer ('working class') residential streets showed one of them completely evacuated despite little serious damage, another heavily bombed but with strikingly more inhabitants staying behind. Seldom less than 15%, often over 50%, of undamaged houses in densely inhabited sectors were abandoned, at least by dusk.

Street-to-street variations became noticeable in Portsmouth. There was no opportunity, under those conditions, fully to study the factors, but we could clearly record that it was so; that if a post-blitz exodus started in one street, it tended to escalate there but not in the next; in that street, equally bombed, things might go on almost as normal.

Much unease, amounting to a sense of isolation, can arise when a neighbour vanishes. The abundance of sailors and dockyard figures – supposedly more 'in the know' than other civilian labourers – was a marked feature of these Portsmouth streets, where 12% of the chief wage-earners were naval, 24% more dockyard. If a Chief Petty Officer or docker's union official went, others could follow rapidly. Moreover, many of the sailors were off at sea: their wives were already alone, with the families.

Occupation of these homes, always rented, showed that 54% of the stay-putters had lived there over 5 years; nearly half of these more than 20 years. For them the wrench of parting was greater: it was not enough, though, to stop them going, if other pressures accumulated.

Getting to know, over months, the dwellers of these streets, the investigator could distinguish a great range of variation. This variation is in constant danger of being lost to view during generalized and abbreviated replays of what occurred. It can become tiresome to reiterate the necessary qualifying phrases. Let a piece from our investigator's notes January 1941 speak to this point:

3. *Individuality*
M . . . Road had the more marked individuality. Among the (staying) people who especially helped to give 'character' to the road were

. . . A plump, white-haired, equable old lady, who wouldn't leave Portsmouth because 'it's our fate', who didn't 'trouble about things; tho' some people can't take it as calmly – they wishes they was like me'.

Husband, a window-cleaner, was not getting much work now, but was 'getting along all right' ... in spite of her optimistic outlook, 'never wanted to see anything like that raid again'.

... A charwoman, wiry, energetic and talkative, who remarked: 'Old 'Itler's not going to drive me out of my 'ome! Everything's ready in *my* shelter! Blast wall, electric lights and all snug! But people ought to be more prepared.'

And of the (last) raid: 'When we got to L ... Road, *there was our place burning*. And the fireman saying, "Who's got the key? Who's got the key?" Lord, I says, Lord, can't you get over the gate? And I gets in and gets the women to work and we soon 'ad the fire out!'

And the other side of the picture:

... A middle-aged woman, nervous and depressed, who said: 'I feel stunned – like – seem to 'ave lost all me power of concentration. It all seems difficult – people can put up with a lot, but there is a limit. I'd make any sacrifice if the war could be over ... there's all the women and children in Germany, when you think about it you don't want there to be so much suffering.'

... A woman of 26, haggard and strained, who had four children under four to look after, and a husband at sea:

'I don't feel I want to get away. You get it anywhere. I feel I'd like a rest sometimes, tho'. I never get away from the children. I get so bored sometimes, I could scream ... I never have any time ... I get terrible tired ... I've never 'ad no real enjoyment.'

And of the raid: 'I don't know how I ever stood it. My God, I never knew how I stood it!'

Remaining people in C ... Road did not stand out so much from their neighbours. There were, however, two elderly spinsters, dressmakers, of better education than the other people in the street, the elder of whom had once been an ardent suffragette, and now seemed to be something of a religious fanatic. Among other things, she remarked: 'I am reluctant to leave my post ... I see a terrible fear on the men's faces when the fire was on ... I look to God Almighty – He will stop the war. We never liked other people's enjoyments much! ... We want to stay in Portsmouth – we've got everything nice!'

And representative of those who showed their depression more clearly: 'I've seen a lot of trouble ... I feel like crying *inside* all the time ... people aren't taking it too well ... a lot of them are nervous! I can't get out because of the baby. And the street's so quiet now. You don't stop people.'[2]

This was, broadly, the variant style from street to street after Portsmouth's first big, January 1941, attack. Despite the measurable inroads of quitting and trekking, the higher police and other supervisory authorities insisted there was very little of either. Many high-ups were so busy with direct administrative problems that they accepted this officially-sanctioned view. Such marked dissociation from a mass trend was made easier by the early destruction of the Civic Centre, less fortunate than Southampton's. The crucial Town Clerk's office (with its associated links to the Regional Commissioner and Regional Information Office), moved into an hotel on the front, physically out of touch, difficult to reach. Police Headquarters went to a former college some distance away. Public Assistance was far off in another building, the Medical Offices at a fourth locality.

This out-of-touchness extended down, more casually, to voluntary leadership and organizations. There was criticism of Portsmouth's churches, no longer seen to be effective since some or all of their buildings had been destroyed by incendiaries – against which they had evidently not been fully organized. At an advertised meeting in January of volunteer fire-watchers in the Co-operative Hall, with only 12 present at the start, 20 later, the speaker described how he'd tried to put out fires he'd found in one church which had 'no hose, no water (buckets) and no sand'. They could have saved it 'instead of being forced to leave the scene' if anyone in charge had provided the minimal equipment. As it was, the church 'became a beacon'.

A special feature at Portsmouth with its very strong naval influence, was the employment of the Army to stand guard through the town against looters. This incensed Army officers (and others): 'As some of them pointed out, *there was no sign of anybody thinking of looting* anything during the raids, and some of these officers felt frustrated at not having done what they considered a useful job in the emergency. This probably biased them into a high degree of criticism of and even contempt for the local authorities; this criticism they were widely expressing and canvassing.' But many functions which the Armed Services could have fulfilled, as at Southampton, fell upon an overloaded police force. Even the old College entrance to the improvised Police Headquarters was obstructed by an enormous safe, TOWN CLERK chalked on one side.

It was the Chief Constable's responsibility, under pre-blitz government plans, to preserve law and order. But the local authority was entitled – indeed half expected – to call in the military. What puzzles one here is the narrow spectrum of authority delegated to the Army. But this is in line with the patently *ad hoc* way the forces were brought in to help, differing strikingly from town to town. No special role was allocated in advance, sometimes no role at all afterwards, despite dire needs. Sometimes one branch took over more or less on their own, as with the RAF in the case of emergency evacuation in Southampton, or some of the transport problems in Plymouth and elsewhere. There was no clear-cut plan; and there was plenty of jealousy.

Earlier, the Mabane Committee, considering manpower resources to combat major attack, had concluded that armed troops would be the only possible reinforcements for over-extended civil defence services. But the underlying fear here was probably of panic and the need to control subsequent mutiny and revolt. The same broad conclusion was separately reached on various grounds in other pre-blitz plans. As official historian O'Brien has remarked: 'It is doubtful whether in any of this planning the full extent of aid necessary from the military authorities after a major attack was really appreciated.'

The same was true in the range of problems where a pre-occupied constabulary could not cope with normal human reactions within the law – as in the influx of sightseers or the predictable urge to 'ask a policeman' stemming from the crippling breakdown of post-blitz information services. The forces could have helped much more in such roles. That they were not so employed was a measure of civic reluctance to admit local 'defeat'.

By and large the January picture of dislocation and distress in Portsmouth with which policemen and everyone else had to grapple was much as in Southampton a month before, so need not be repeated here. In mid-January about the most conspicuous aspect of otherwise drastically depleted shops was the abundance – 'in some cases practically the only feature' – of pheasants, selling round 19/- (95p) the brace.

By contrast, the town's two communal feeding centres, inadequate and still more inadequately publicized, were offering:

Tea – ½d per cup
Cocoa – ½d per cup
Soup – 1d per cup
Minced beef and potatoes – 3d per small plate,
4d per large plate.

At those rates they sold out of solid food soon after 1 p.m., with customers sitting on the benches, the floor, the stairs and in the winter street outside (18 January).

Closing note of a Portsmouth shaken by the first and lesser of its three big attacks was a mass burial, on 17 January, of some of the blitzed dead, including a man of 92, two brothers, two married couples, a mother and daughter, attended by some 50 mourners (and 'in their midst an investigator'). The affair had been advertised and treated in editorials in the local press. It was graced by the presence of the regalia-laden Lord Mayor, Town Clerk, Anglican (in blue and gold vestments, with tall mitre) and Catholic (cherry pink, black and gold with skull-cap) Bishops, lesser clergy (the only time most of these were observed in January), City Councillors, Naval and other officers.

The long procession crossed the afternoon town from Alexandra Park to Kingston Cemetery, led by the Royal Marine Band, trombones emphatic, drums muffled in black crape. Behind came a top undertaker in a top hat, followed by twelve Daimler and Rolls Royce hearses with top-hatted drivers and coffins draped in Union Jacks. The armed services walked alongside, acting as pall-bearers, at start and finish in slow march, with a Naval Commander in charge of the whole parade, including French Navy, RAF, AFS, ARP, Home Guard, etc.

Spectators, police-controlled, lined the long route, crowding the pavements in places. Many dabbed their eyes, some women (only) bit lips or knuckles. Uniformed onlookers saluted the hearses, some keeping that position until the whole procession had passed. So did one postman. Many stayed on long after the cortege had vanished, discussing what they had seen. Mixed with reverence and tears was, of course, curiosity:

'Get to the front, duckie, we might be able to get in.'
'I'd love to see the grave.'

That grave, a hundred feet long trench, was lined with Union Jacks. At one end, the fifty-odd mourners; at the other, fifty mixed clergy, led by those bishops.[3]

Not surprisingly, this masochism *en masse* registered 'a strongly depressing effect on the town just a week after a big raid, when people were beginning to recover their good spirits'. It was nevertheless the most conspicuous single manifestation of leadership in Portsmouth during the month that elapsed between the two biggest raids. The second, considerably heavier, occurred on Monday, 10 March, after quite a noisy Sunday night too, before a burst of beautiful spring weather. Portsmouth folk felt they were raided seriously three nights running, though that weekend was quantitatively no more than a 'normal' run-up to the larger climax. These smaller raids leading to a big one were probably more significant in causing a build-up of anxiety without leaving enough time to intensify 'night training' in the London style; but individual reactions varied greatly.

The heavy 10 March attack concentrated rather more on the dockyard than before, though still heavily hitting the civilian city. There was, however, much verbal exaggeration of damage to naval property. People wanted to believe that it was very grave; then (as in Southampton) thinking there would be less reason for the Germans to return. By now, too, townspeople were more used to their blitzed scene, especially in the centre. The new big raid, linked to intervening smaller ones, was not so visually upsetting.

Yet many considered the March blitz impact sharper, 'worse'. More heavy bombs and greater noise were repeatedly stressed in conversation. Some felt amazed, many surprised, that anything worse than January was possible. They had largely assured themselves of their own courageous, unshakeable response to anything on the scale of Coventry.

What mattered more was the preparedness of Portsmouth as a social-civic unit. This time it did not look so good to the remaining civil population, who had been expecting, vaguely, that in the two months since 10 January local arrangements would have been appreciably improved. This time, even the AFS, praised in January, came under hostile comment. There was much talk that fire engines coming from as far as Outer

London could not fit their pumps to the local hydrants and in consequence sat unused. Other comment was on familiar lines. A strong minority, not heard elsewhere so far, were openly critical of the local authority, especially as regards feeding (there were no visible mobile canteens, for instance), evacuation arrangements, the locking of churches and church halls. This last had by now become something of a Portsmouth custom, extending into the countryside, supposedly to keep trekkers and runaways away. The clerical attitude led to some bitter reproaches, charges of 'unchristian' conduct and the like.

Portsmouth people now sounded much less sure of themselves than in January. Talking again to the same individuals, one 'noted a significant change'. A number of those who had said they felt fine before – and were liable to be scornful of other, weaker brethren – now openly admitted feeling bad themselves. Signs of open 'defeatism' were few, but more women especially expressed the wish that the war would end soon, they couldn't stand much more. For the first time after a raid, a wall-chalking shouted: STOP THE WAR.

This mood was confirmed by social workers, local officials and a town councillor. The Lord Mayor, however, retained a hundred per cent rosy view – insofar as anyone from outside could penetrate that piece of the civic façade. He and the Lady Mayoress were seen to be energetic, devoted, 'hard at work smiling at everything'.

The trend to trek or quit further gathered force. Bus services outward were saturated by teatime. Cars continued aggravatingly 'unhelpful'. Of 108 private cars now passing the bus queue in peak exodus time, half had at least 3 empty seats, only 5 stopped to pick up a few of the waiting thousands.

On 12–14 March, households in our street study were checked to find how many of the previous stayers continued *in situ*. More than a quarter of those revisited homes were still in streets without serious damage:

- 8% of houses occupied on 1 March were totally abandoned by 14 March;
- 13% of families still sleeping at home then were now newly sleeping outside the city (mostly trekking);

– 35% more said they would evacuate if circumstances permitted (and some later did: see below).

Altogether, 28% of these streets' inhabitants had gone between January and March, 21% more went after 10 March. 'Would you if you could?' had been put to them a month before. Among those who had said *no*, then, there was no significant change. The yes-people who remained often said they would like to go for a spell, rather than permanently. 'I do feel like a rest.' At the other end of the scale were the quarter who spoke up toughly, with an in between 'wait and see' group.

The wives of each stay-put family gave their reasons, in this order of frequency:

1. (far and away first) Husband's work/welfare
2. 'Fatalistic' acceptance of war conditions
3. Dislike of losing domestic independence
4. Expense of moving; obligation to pay rent
5. Inability to find suitable alternative
6. 'Cowardly' to leave
7. Religious or related principles ('I trust in God')
8. No clear reason: just stayed ('I haven't thought about it really').

Number 1, bread-winner solidarity, was indeed solidly feminine: 'They don't look after themselves if you leave them.' Some wives were amazed at the mere idea. 'Me (leave)! With *my* family? Not on your life!' The fatalists clearly cut across other categories, too. Most simply: 'I wouldn't go. What's the use? If anything happens to us it's our fate.' Or: 'What's the use? You'll get it *anywhere*.'

Whatever the stated reason, stayers of any sort felt free to criticize goers. 42% of them blamed the absentees for irresponsibility and the like, another 12% made milder comments: 'I don't think anyone who 'asn't lost 'is 'ome should be allowed to go. It's not fair.' Or: 'I don't think it's fair. I wouldn't go – not while I 'ad me own fire.'

Particularly remarked was the danger from incendiary bombs in empty houses.

Whatever everyone said and thought (not necessarily thinking what they said, or saying what they thought), much of Portsmouth's population was sleeping away from home by the second half of March, and many more were ready to go, while the authorities, local and national, went on treating this situation as if it did not exist. Seen at this stage: 'the situation in the surrounding villages ... is utter chaos.' A Chichester shop-keeper put it: 'If you want to find a needle in a haystack, then go out and look for a room.' Our report of 15 March gives an impression of the city as a whole, reeling from the 238-plane attack of 5 nights before:

Everything seems to be left to the individual. Really *nothing* is being done in Portsmouth to make ordinary people rally and consolidate and gather together the emotional shreds. Moreover, the evacuation of women and children is handled in the extraordinary manner which demands that the women must make their own billeting arrangements in advance, and prove that they have billets arranged before they leave, if they are to get billeting allowances. Owing to the already congested nature of the area, which was allowed to become congested by individual evacuation without any central plan, many women are unable to satisfy this condition, while others misunderstand it owing to the general and continuing ignorance about social facilities and regulations.

These 'helpless' women spoke out unafraid, loud-voiced, unashamed (most readily to other women):

'*They ought to do something to help.* We can't stand any more of this. We'd like to get away but you can't afford to be away for long. They sent us away for a fortnight. But we had to pay 17/6 – you can't pay that and pay your rent too.'

'I'd give *anything* to get away. But *they won't do anything. And I can't afford it unless I get the 5/-. Makes you feel quite hopeless.*'

'My God. I wish I could get away. Look, I've got four children, all under four. I went up the place yesterday and they say they can't get me away unless I can find somebody to say they'll have me. I burst into tears – couldn't help it. I don't know anybody to go to, and I can't find anybody. *I feel as if I'm going mad. They ought to help you.*'

Such quotations could be multiplied many times.

Put crudely, not more than half, and in places less than a third, of Portsmouth's supposedly resident population was still really living there by the end of March 1941. In the parish of Old Portsmouth, at one point only 250 could be located out of some 5000. Some streets there were 'utterly devoid of life'. At night the whole town was 'a tomb of darkness', without social life, though irregularly interrupted at times by the cry of loudspeaker vans, those ravens of post-blitz. For when the air-raid siren system broke down, as it did in Portsmouth, the vans were used as warning substitutes (and for little else).

Loudspeaker vans and other daylight sources of emergency information were never more needed, to speak into the characteristically echoing cavern of post-blitz confusion and misunderstanding. Monotonously, as before and after, lack of information was 'a major inadequacy'. Local tests showed this confusion among police and wardens as well as the civilian mass. For example, hardly anyone knew how to direct you to the one Citizens' Advice Bureau ('in a town which requires at least five'). Those who managed to reach this reservoir – or oasis – of verbal help found a handwritten notice of identification outside; inside one or both of the Misses Kelly, intelligent, thorough, coping with questions but (they said) experiencing 'the greatest difficulty' in themselves getting a necessary minimum of information.

The contribution of this modest semi-voluntary advice bureau was great in the post-blitz period. In 1940 it had processed 4455 enquiries – 900 on air-raid damage, 392 on shelters and gas masks, no less than 453 about the Channel Islands and 43 more about overseas evacuation for children. The value of such centralized information sources – not built into the central civic core but operating independently on behalf of it – was very slowly realized in most places. In London they had not existed when the blitz began; by May 1941 there were 78 information centres for the 96 local authorities in the London region.

In Portsmouth, even the key Local Information Committee could barely be described as better than stagnant. An enquiring observer was 'regaled with anecdotes from various members separately running down each other collectively'. The appointed secretary already had many other vital air-raid and recovery duties. A tendency to pile important jobs on a few local leaders was, however, no stronger in Portsmouth than elsewhere. It was a general trend; and, in blitz conditions, a most unfortunate one, since a man (perhaps of no great talent) might suddenly be faced with trying to do several jobs, each of which needed a superman.

The unflinching desire of civic leaders, elected and appointed alike, in those days seemed to be to distribute power – especially new power – among themselves; failing that, within their narrow elite. It was not thought of as a time for new men (and women) to implement new measures.

Over this uncertain and in a sense divided coastal city fell their third and last big night, 17 April, with a record 249 bombers. A smaller but still heavy raid on 27 April, with 38 planes, rounded off the extending suffering (particularly hitting one of our study streets in North End) with 22 landmines. This last attack also demolished much of the Royal Hospital, an important surviving landmark.

Not until the next month, on 14 May 1941 – four days after the last big attack on London – did the *Portsmouth Evening News*, more receptive to voicing criticism than most wartime papers, give an account of a City Council meeting including this:

A voluntary Evacuation Scheme
The Lord Mayor reported that as a result of a tour that he and the Lady Mayoress made of the shelters in the City and on the outskirts, he saw the Regional Commissioner and said that *something would have to be done in the way of an organized voluntary evacuation* of the women and children ... he ascertained that about 75% of the women and children he met would be willing to leave the City and husbands would be willing for them to go.

75% of those ready to speak out to the Lord Mayor left precious few true stay-putters by spring; the figure is higher than any of ours. But this Lord Mayor was no alarmist – rather a conservative loyalist now very belatedly alerted.

The Regional Commissioner took up the evacuation of women and children with the Ministry of Health, until then '*dead opposed* to any voluntary evacuation scheme'. Lord Chatfield, a distinguished Admiral, was especially appointed to go into the question. The Lord Mayor said that he had 'no shadow of doubt' a new scheme would emerge.

On 10 June, however, at a further council meeting, several councillors were strongly critical of inaction. One wished 'to draw attention to the *preposterous* state of affairs which still existed in Portsmouth after many months', notably in providing evacuation billets for children. Their comments were talked out.

Many small children had in fact left Portsmouth long before. To the visiting eye, by spring it was a childless city. Those that remained showed many signs of good adjustment. Only 1% of the households involved in our street study expressed concern for the welfare and development of children except insofar as: 'they're sure to be nervous.' One mother proudly described her son, 4, counting out falling bombs, although 'the little girl next door got hysterical.' The youngest seemed to mind least, usually: 'They sings and laughs in the shelter. They don't mind a bit.' 'The children who've been evacuated and *come back* are more nervous than the ones who've never been away.' The children who did stay, by all accounts enjoyed life, shelter life included, though one mother reported: 'My boy cries if I take him to the shelter at night.'

Adults, likewise, had very varied views on shelters. In the four study streets, sheltering came high but well behind health and home economics when housewives were asked to name their main domestic difficulties, which they rated thus in approximate order of alleged difficulty:

1. Health and general strain ('nerves', etc.)
2. Money worries
3. Food supplies and rationing (overlapping 2)
4. Shelters
5. Fuel and cooking (especially erratic gas pressures)
6. Broken windows and other minor house repairs.

Many of those difficulties arose from war itself, raids apart. And a large group mildly said that they had *no* outstanding difficulties worth mentioning. Although most of these women went on to name something troublesome in discussion, a hard core of true 'fatalists' took the view that: 'nobody can do anything really', so why worry? Or 'I don't worry about anything.'

Shelters were not so great a concern, partly because people felt this was something they could no longer do anything about except in minor respects and because at first, at any rate, the system had satisfied a fairly high proportion of Portsmouth's remaining population. Thus (after January) 41% were prepared to accept that their available shelters were good against 17% who thought otherwise and a non-committal 42% which included those not using them and many who took them for granted anyhow. More than a third (37%) of those working-class housewives who stayed behind admitted to sheltering regularly on alerts by day in the spring of 1941; 63% did not, though many of the former qualified their statement with 'if there is any gunfire' and the like; and the latter must be viewed with some scepticism (the difference between 'say' and 'do' where 'bravery' comes in). Some sheltered if away from home, where they preferred to remain indoors: the fear was more of falling shrapnel than of bombs. An observer noticed that nearly everyone in one street had come out to their front-door porches to discuss whether 'anything was going to happen' at one day alert. Then – 'as if by general consent' – they moved off in a bunch to the entrance of the public shelter on the corner pavement. One old lady who had earlier been emphatic that she *never* left the house for a day raid, went with them. Although there was no gunfire, the group remained beside the shelter until the all-clear 15 minutes later. Whatever the uncertainties, only 2% would agree that they went to day shelters more since the first big January attack, as against 93% who said they didn't (much as in London earlier).

Night told another tale. Only 18% then stuck to their homes: a few because they had no shelter readily available, most because (they said) they actually felt safer indoors. 27% normally slept in their own domestic shelters regularly, and 30% went regularly to public shelters if there was a warning. Another 23% waited to see how things developed before sheltering – these being mostly families with their own garden or backyard accommodation nearby.

Public shelters were usually *full* soon after night alerts, so that waiting-to-see could leave one out in the cold. These public shelters were mostly brick structures built on pavements and in empty lots, large boxes with slots for bunks, though generally at first without the bunks themselves. Some were only one brick thick in places. Customers noticed this, although more or less accepting them as the best they could get. Solicited opinions include that of the wife of a City Corporation labourer who quoted her husband as saying they were not properly finished – 'and 'e ought to know'. A brickyard labourer declared the materials used from his yard were 'a lot of rubbish'. They were working in a terrific rush to get something up earlier on, night and day for 6 weeks (£8 a week overtime wages). Quality suffered.

Long after the work-rush had passed, complaints on the score of quality continued up to City Hall level. There, in May, three councillors voiced complaints with force: bunks available but not installed, standing water, cement crumbling into sand (a bag of it exhibited) and so on. The same Alderman who had killed off the evacuation amendment brushed off this one also with a brusque: 'application should be made through the warden.'[4]

A public shelter commonly served several streets (in practice an average of roughly 150 families). Some were designed to take nearer 50, though the exact quantitative intention was seldom stated and did not appear to be based on close study of actual requirements. In one shelter, said to be for 50 persons, an evening headcount gave 103. But overcrowding on this scale was not necessarily disliked. Women, mainly, thought it 'better with a lot of people', and 'I like to be with plenty of people.'

The population of each Portsmouth shelter nearly always built up into a steady pattern of 'regulars', reminiscent of the public bar in a local pub in the same sort of district in peacetime. Whereas daytime usage went down a bit later in 1941, night-time usage stayed high. The 'publics' normally had little or no room for 'outsiders' – and could not have coped with nearly the whole pre-blitz population of the city. On one occasion the regular pattern was sadly disturbed when incendiary bombs fell on a piece of wasteland holding shelters. An excitable warden ordered everyone out: 'A panic followed as they rushed away and tried to push into already overcrowded shelters', kindling resistant fury within.

Private shelterers overwhelmingly used the Andersons in their gardens (usually tiny), with a small number of more solid family brick shelters, numbers of which were built early that summer, with overhung concrete roofing and a tarred layer. A high proportion of private shelters stayed somewhere between damp with sporadic floods and a semi-permanent water-logged condition. In parts of Portsmouth the Andersons were pretty well unusable: 'I've got an Anderson and so have other people in this road. It's half full of water like the rest, and we can't use them. I've written twice about them to the Council, and nothing's been done ... There aren't enough public shelters round here. Lots of people say so. I say it's not right. Something ought to be done for the people who've *got* to stay. We're only *human*, for all we're willing to put up with.' And: 'My shelter's damp and so is everyone's. Something ought to be done about that. You don't feel safe.' An unexpected snag arose in more crowded streets where, it was found, the pipes from other houses converged under every sixth garden or yard, making it impossible to dig deep enough for an Anderson-type shelter. An old woman so affected was bitter that 'they' ought to do something 'where the pipes come through'. Mostly, these 'sixthers' did not know about the problem: 'We've no shelter. I don't know why.'

The main trouble, however, was with bare earth floors. Few seemed to realize that there were ways of combating flood and similar troubles. Only occasionally did we find anyone in Portsmouth who 'took pride in the upkeep of the shelter': by painting it inside, putting linoleum on the floor, tarring over any sort of flooring, laying an old rug. 'Blast wall, electric plug and all snug', was a rarity.

Although there was quite a lot of irritation at the shelter position, only a small minority were at all deeply concerned. This should be seen against the return in the summer of man's innate optimism, that almost tireless capacity to look past death and foresee brighter days. Put in mundane terms, in mid-1941 Portsmouth, for our study streets:

- 55% definitely did not expect any more heavy raids (July; up from 2% in February).
- 14% did expect (incorrectly) at least one more heavy raid, compared with 39%, still a low figure, who so predicted (correctly) in February.
- 54% believed the war would end during 1941 (down only 10% since February).
- only 3% said they would evacuate if circumstances permitted (a drop from 35%).

All these figures differently reflect a high rise in Portsmouth's self-assurance that summer. Nevertheless, a big slice of the population stayed away; while an eighth of those remaining expected more attacks. At that stage, a third of those staying in town *still* had no private shelters. The public ones were, by July, considered good enough by a third, good, though with qualifications, by a fifth, decidedly bad by only one in 17.

In addition, a plan to bore *deep* shelters (non-existent in Southampton) through Portsdown Hill three miles out, had been initiated and tunnelling begun. Workmen engaged on this project spread the encouraging word that the complex would be elaborate: there would be rooms for families, dormitories for singles, hot and cold water, reserve oxygen, attendant doctor and nurse – all to be ready by October 1941.

There was much talk around Portsmouth that summer about deep shelters, which had proved their worth in London the previous winter. The so-called Kearney Plan, put up by an enterprising local engineer, aimed to provide a huge deep system under the city. This was also the subject of controversy in the local press, starting on 17 June 1941. A report favouring this plan, however, was rejected by the City Council, though not without public accusations of graft and prejudice.

Interest in these matters did stem from a deeper anxiety. However much one said the raiders would not return – and the act of saying so could, like prayer, seem to help – what if they did? There was at that time a very widespread feeling that 'another winter of the same' was not to be endured with the private and public shelters so far available. Reflecting this underlying anxiety, some who on the surface said all was well would nevertheless say when asked about the Portsdown Hill and Kearney projects: 'They ought to have done it all before. They knew we was going to have war didn't they? They could 'ave made an underground city of it.'

And, half-correctly: 'They'll never be finished in time for anybody to use them. It takes a dictator to get things done quickly.' Dictatorship was by no means involved in this or any other British deep-shelter planning. The official concept, from back in peace, was (as we have seen) to keep the civil population 'dispersed', in the belief that this would lessen vulnerability from the patterns of bombing, as then foreseen. Government did not want a lot of people sheltering in any one place. Had there been deep shelters prepared by October 1940 in cities like Portsmouth, a great deal of trekking and other dislocation would probably have been avoided – with consequent savings in distress worth far more than the cost (maximum £25 a head) of the necessary work.

There had been related changes of mood as air attacks began to phase out over Portsmouth, starting after the big raid in mid-March:

- 68% then said they had no time for leisure pleasures outside the home at present (itself a fall from 86% in February; to go down to a mere 14% by July).
- radio was felt to be the biggest remaining home pleasure (many sets still used batteries).[5]
- 5% wanted to go on living in the same house, although seldom satisfied with these as final homes.
- yet 61% wanted to remain in post-war Portsmouth.
- 91% said they preferred living in some sort of house rather than a flat (although there was a general expectation that rebuilding would be largely of the latter).
- a mere 7% of family wage-earners had lost their jobs or steady work through the raids.
- and three-quarters of these last were re-employed within a week or so, locally.

These figures, crude as they are, do usefully reflect the basic instability of urban life in those days. It was this which made quitting easy whenever other things were not too difficult. Money worries and most directly the wage-earner's essential need to stay at or near the job, were the prime considerations which kept householders on the spot or – failing that – trekking to and fro. Without these economic sanctions, it is possible that nearer 100% might have quit.

The figures for lost jobs – actually 6.8% – is perhaps the most important from the war-winning point of view. Disregarding all the human decencies the principles of democratic sharing and such like, work was what would eventually count on the home-front. The failure of heavy air attacks lastingly to dislocate industry and the general war effort in places like Portsmouth was to emerge, after the war especially, as one of the more marked non-results of the whole process.

Back in February, after the January blitz, workers at Portsmouth's big naval dockyard were showing the effect of 'long hours of work under war conditions; with the strain reaching into their families'. Conversely, though – and characteristically – about a quarter of these families were sleeping away from Portsmouth at nights and weekends. But in few of these families was the docker doing any less work than his stay-put colleague. Very few dockers quit totally.[6]

In July 1941, a repeat study showed a small drop in dockyard trekking, from 27 to 21%. At peak, around 10,000 men were involved in this work. Their post-blitz performances were high. For reasons beyond worker control, work slackened that summer. Criticism of old-fashioned naval procedures became common, in fact:

'They still think they're back in Nelson's time.'

'They think because Nelson did a thing, they've got to do it too.'

But that had been half-true all through the city for months past: old-fashioned answers to new problems.

The Portsmouth authorities were certainly slow – though in no way uniquely so – in taking effective action on evacuation (even when, eventually, supported by the Lord Mayor), recognition of trekking, shelter needs, and so on. In this they cannot be said to have lagged far behind their citizens, only to have failed to lead them into a faster tempo. Thus between raids in March–April, 13% of those remaining in town said they had made new preparations for raids since the last blitz, whereas 19% flatly stated they had done nothing; the rest, two-thirds, belonged to that silent majority, the don't knows. This had to do, of course, with the reluctance to foresee further raiding (39% in February), which, as an investigator then noted – not without subjective irritation – 'was a question people did not wish to discuss' any longer.

In writing about past situations like Portsmouth's towards

mid-1941 or Southampton's late in 1940, one has to distinguish between leaders and led: those whose duty it was to decide matters affecting many, and those who seldom decided for more than their families or themselves within a framework established from above. But it cannot be over-emphasized that local leaders were subject to the same pressures, fears, anxieties, wishful thoughts, as those they led. They were motivated by ordinary jealousies, ambitions, pride – most relevant here because this led to 'poohbahism'; the concentration of key jobs, some new, in a few established hands. More important, perhaps, they were 'ordinary' citizens too, as well as extraordinary. No one, whatever his status or responsibilities, could escape the pressures of the blitz.

So some top men in key posts reacted like trekkers, others like stay-putters. A few nearly panicked, a few barely noticed the chaos; most moved between any extremes. Plenty of key officials kept going, but with a low threshold of confidence in facing the facts, now shattered out of old patterns. Top men could crack inwardly, quit intellectually (as in Southampton), while clinging to the robes of office, maybe hiding behind them. No case exists, so far as we know, in which a provincial leader publicly admitted inadequacy and so resigned.

Very few females had the opportunity to get into a position where there was anything for them to resign from! Looking back from these Women's Lib seventies, the failure to use women in certain blitz situations seems astonishing. The feminine role in dealing with civilian and especially household problems emerges as grotesquely small. Matters of war were still planned by and for the male. Rare were the Misses Kelly, exercising influence in the void caused by failures of communication; the female voice speaking domestic words to the distressed cook, mother, shopper, home-sitter. Plymouth had its Lady Astor, indomitably feminist. Portsmouth had not even the ghostly vision of a Lady Hamilton.

(3) Three 'defeats'?

Basil Collier, looking at the tactical angles of Civil Defence as a whole, termed 7 September 1940, the start of the London blitz, 'a victory for the German bombers' – because the R A F could not prevent the Luftwaffe attacking 'for seven hours ... over London unimpeded', although balloons and guns prevented them from closing to point-blank range. (The guns numbered 264 for the zone, 216 less than the number officially considered minimal for London's defence by the pre-war Committee of Imperial Defence.)

The Luftwaffe often again thereafter had seven hours which, though impeded, involved them in little or no aircraft loss. But the premise for victory must surely be the achieving not just of a theoretical supremacy in weapons but achieving success by results: that is, by hitting the targets with bombs or causing other damage. The Germans, limited after 7 September to mass attack only by night, could not hit targets more specific than industrial or (especially) dock areas easily identifiable from the air. The major bomb-load fell, of course, on the civilians. This being so, defeat or victory depended primarily on 'success' so far as those people and their environment were concerned.

As regards environment, the results were there for all to see, though difficult to measure. The material loss was obvious enough, but in London never decisive. For the rest, the people, it is clear that German policy did not succeed. There was no 'defeat' for the Londoners who stayed; the ones the journalists and commentators saw. London could and did take it. And that was a victory for Britain, especially if we leave aside those who left town or otherwise opted – often for excellent reasons – out of the battle. It was also an encouraging example to Britons and other friends throughout the world. It demonstrated how ordinary people could carry on. But examining exactly what happened today, there is no need to adopt the exaggerations, the propaganda of herosim, which was lavished upon the Cockney with the laudable purpose of encouraging him and of discouraging the enemy from continuing to attack him.

Anyway, by mid-November 1940 the Germans themselves had recognized that 'London' as the target of their nearly total effort, did not look like being decisively 'knocked out', let alone Britain forced to capitulate. They decided, therefore, to widen their targets – as stated in the quotation from Collier which heads this chapter. Failing to follow up Coventry, thereafter they concentrated on a few large inland cities and – much more – on easily identifiable coastal towns. Those on the south coast, like Southampton and Portsmouth, though not so important as the Atlantic ports, could be approached most rapidly and suddenly, without crossing British soil and ground defences. The German targets in all these places were only nominally pinpointed. The idea was still to dislocate the whole area, by shattering social and domestic patterns at least as much as anything directly industrial or military; in the ports, naval and dock personnel supplied an added attraction.

The fact needs to be faced that once the bombers switched to smaller places and to highly erratic patterns of attack, they had more civilian success. Coventry, Southampton and Portsmouth; each in its separate fashion represented a sort of defeat on the ground; Coventry, indeed, was a one-night rout. Southampton and to a lesser extent Portsmouth were battles of attrition, wearing populations down into large scale though only partial withdrawal. And, as the dictionary reminds one, defeat has important secondary meanings: to *frustrate*, to *baffle*. As much as anything, Sotonians were defeated in bafflement.

Within these overall setbacks on the home front, two levels of defeats can reasonably be distinguished. For whereas a normal, regular uniformed army stands or falls as an entity and the defeat of the officers without the men is practically inconceivable, in this style of war the mass of 'other ranks' could – and definitely did – operate without or against orders, ignoring by dereliction the few instructions audible from the top brass above: from Town Hall or Whitehall. In these three provincial places, each of medium size, the 'general staff' on the spot were more markedly defeated than their other ranks, the 'lower orders' whom they failed adequately to command; and who, leaderless in the face of hostile events, withdrew to unprepared mental and/or physical positions, including (if they saw fit) positions as far as practicable removed from the scene of the battle. 'Trekking', which government from top to bottom ignored for half a year, was the simplest expression of this withdrawal: a form, superficially, of defeatism. Yet the trekkers kept on working: and kept the peace of the realm. By so doing, regardless of the battlefield, they defeated their enemy's prime object in making these costly attacks. At the same time, by so continuing to maintain the economy, the war-machine of industry, commerce and the ports, these other ranks achieved, against all odds, a sort of victory; not least a victory over their own higher brethren, the planners and politicians who had expected them to panic, to need troop-control, to go gibbering to the sick wards or to cower as shelter troglodytes.

Presumably it can always be the little things that defeat any force. An army marches on its belly, for example. No one seems to have applied this elementary reasoning to towns under air attack. The same thinking which failed to foresee the need for factual information (not propaganda or 'morale boosting') at the street-leadership level equally failed to bother about the little problems, so easily solved, so vital to many. It is hardly too much to say that a good many abandoned Portsmouth because they could get a fag and a pint out Chichester way.

Trekking is emphasized here as the most visible expression of a civilian reaction which combined rejection (physical defeat) with specific determination to respond to reality, to live not so much through as around the blitz. London's huge size and alternative outlets allowed for similar but less obvious reactions, not necessarily requiring such drastic spatial severance. The effect was heightened in the coastal towns because they had only half of a circle to move in or across, whereas inland targets offered 360° of escape. It is doubtful if the Germans recognized this considerable incidental bonus on port bombing, any more than they saw the profit in stepped-up incendiary raiding at week-ends when (until well on in 1941) voluntary fire-watching tended to be slacker in public and civic buildings, churches and even factories.

At another level, if trekkers and full evacuees are completely separated from stay-putters, the picture of 'defeat' alters. Nevertheless the idea – which has largely survived into war folklore – of the stay-putters as happy warriors, especially brave and cheerful, is inadequate. How could one distinguish the degrees of intelligence and stupidity, relief and anxiety, pleasure and misery, inside each human equation, whether staying put or flitting? So much depended on personal temperament, family organization, home ownership and the like. Refresh the search for generalization by glancing back for a minute at a diarist living on the outskirts of Birmingham. On the night of 14 November, the coventrators flew (she felt) 'in droves right over our roof top, hour after hour'. Thoughts centre on her sick daughter (age 11) for whose sake they do not wish to shelter in their chronically damp Anderson: 'We were forced to stay indoors because of the hopelessness of the shelter. There I sat on the bed on the floor in the corner of the big room, ready to fall on top of Jacqueline the minute I heard a bomb whizzing, wishing I had moved our heavy gramophone from overhead, wondering how big a dent it would make in me.' Five nights later Birmingham itself, with diarist, was blitzed:

Oh God, what a night.

10.3/4 hours of anguish, misery, hunger and sleeplessness. 5.50 p.m. Tuesday I had cooked a dinner and was about to serve it up when the sirens went.

Les (husband) came in and found Jacq and I sitting in the cubby hole. I said 'I'll just dish the dinner up when the planes have slacked off for a moment.' They didn't slack off, they got worse, a terrific bombardment started and then the house shook with bombs exploding.

Les said 'Dinner or no dinner we're going down to the shelter, I hope you've bailed it out.' We scrambled into our clothes, carried a blanket and cushions and hoped for the best. The sky was one blaze of light, gunfire was going on and did we run.

We all got in safely, the floor was one large puddle, we sorted ourselves out and made the best of it, till a lull came. No lull came, as one wave of planes came over and was dying away so the next came into hearing, every 3 minutes I should think.

About 2.30 a.m. beginning to acclimatize, she came up to finish readying the dinner. But just look at the suffering, where no bomb fell anywhere near; the extent to which domestic improvisation and nerves had to come into play to decide what to do, night by night. To these multiple difficulties was superadded always in the provinces the erraticism of attack, the seemingly 'senseless' pattern of German blitzery. This made learning a routine and keeping to it that much more difficult. It was something like going to school on Tuesday one week, Saturday and Sunday the next, then not at all for a fortnight! More 'fun' that way, perhaps; but hard to succeed as a scholar.

With these thoughts in mind, tentative and provisional, the record for the other main southern ports may be seen more rightly in perspective.

(4) Bristol, in the South-west

About 400,000 Bristolians (twice the number of Sotonians) received six full blitzes with a nominal 919 tons of HE (to Southampton's 647) between 24 November 1940 and 11 April 1941, a wide scatter pattern with one semi-doubled assault only: 3 January on the main city, next night mostly on Avonmouth docks.[1]

Whereas we kept studying Portsmouth and Southampton long after the last big raid, in Bristol we started long before the first. A resident whole-time observer was so blasé by 27 August 1940 that she reported that 'she did not propose to write reports on raids happening *at night* every time but will report on day raids (unless you let me know you particularly want night ones)'. Ten days before London and ninety before its own first great test, Bristol already thought that anything in the air was an 'air-raid'. To some extent this early over-reaction helped instil sound raid habits. Offsetting this, many were misled into underestimating the trouble to come, so under-reacting to precautionary preparations, especially at the level of upper leadership. Our observer's own pre-sophistication was triggered by a raid the previous night, 26 August, with sirens 10.05 p.m. and 5 a.m., during which only a young married couple moved down, with pillows, to a cupboard just deep enough to take prostrate bodies, the house's only shelter. The landlord informed his lodgers that his wife and he left the cupboard free, as they 'go down to t'kitchen in bad raid'.

This pseudo-conditioning was probably a misleading factor in nearly all provincial centres. It became striking, at Bristol, when it continued, complacently, even *after* Coventry (14 November). Public shelters were nearly empty except when the aerial activity was loud. Stray day warnings did close some shops. When Filton air works were raided in September, a labourer reported on his mate up the ladder: 'He didn't bother to come down ... he jumped.' That day, too, three men discussed a German aviator who baled out at nearby Shirehampton:

'Before he was down he got out his tommy gun and shot at two farmers. And what do you think, some of the women there brought him tea and cigarettes – I ask you, tea and cigarettes.'
'*I'd* give him tea and cigarettes.'
'Fancy that, giving him tea and cigarettes.'
'One of the farmers had a go at him with a lump of wood.'

Thus Bristol moved into the war's second winter looking publicly prepared. Even those who did dread heavier attacks mostly said it was not going to be as bad as they had earlier thought; and, especially, that preparations were adequate. It seemed as if things would work out quite well under pressure, although the cheerful failure to learn from the previous experience of others was somewhat disturbing.

The big bombing of 24 November was, as Angus Calder has remarked, 'something quite new' for Bristol. But even after the second, on 2 December, M-O's blitz team noted: 'Damage in Bristol is considerably less from the civilian and town (centre) destruction point of view, than in Coventry and Southampton.'[2] Public utilities, for instance, were working well, the telephone system practically normal. 'There is' (we wrote then) '*nothing like* the dislocation of everyday life and the *multiplied personal discomfort* which still dominates the whole of life in Southampton.' Hot meals for the homeless were quickly available, for the first time in a provincial blitz. And it was with some surprise that one found a 'remarkably low degree of private evacuation and desertion ... The working classes, in particular, have overwhelmingly "stayed put" – unlike Southampton or Coventry.' The Bristol countryside was already full of London and other blitz evacuees, favouring the west country (but not too far west). The local Ministry of Health office reported not less than 200,000 outsiders entering the region in September and October 1940.

This sunny mood was not fully to survive the winter, but Bristol started with and retained two big advantages. First, it was a large conurbation, much less concentrated than the other southern coastal towns, its resident population exceeding Southampton and Plymouth together. Second, and at least as important, it was the designated centre for the 'Region', which meant that the Regional Commissioner and related main officers representing Whitehall in the emergency set-up were on the spot to react quickly inside the local lines of intercommunication, established in preceding months largely through personal contact – and alive before the big events. Bristol was the only smaller city which had this advantage of the RC within the community – far away Plymouth, for example, was also administered (or more precisely overseen) by this same supremo, while other regional centres were in the great places like Manchester and Birmingham or out-of-the-way, as at Reading and rural Tunbridge Wells. Even so, even in Bristol, intervention was very cautious. As Terence O'Brien emphasizes, 'no Commissioner *ever* exercised his full powers'. Not more than one tenth of such powers might be nearer the average.[3]

Alongside these advantages, Bristol had a strong though by no means unique weakness: a public shelter system felt by many locally to be inadequate. This feeling was aggravated by shelters being demolished and rebuilt, in public view, including many brick and cement shelters which had been insufficiently mortared in the first instance. To quote a mid-December report, written before the heaviest raids: 'There is a violent minority dissatisfaction with Bristol shelters, and this is certainly often spontaneous, non-political, and actually justified. Investigators with a wide comparison of experience with town shelter facilities consider these in Bristol to be strikingly inferior and inadequate in many parts of the town.' An offshoot of this dissatisfaction was a demand for access to seven underground tunnels, supposedly suitable for immediate use, except that two of them were said to house museum and art treasures. One of these was in effect 'taken over' by about a thousand people at this stage, despite – and in part because of – rumours that the BBC were seeking it for their own purposes. By mid-December half this tunnel had rough bunks; half was flooded. Conditions in the used part were similar to those 'in the East End at its worst period', according to observers who had seen both occupations. Every foot of dry or dryish space was taken up before blackout every winter night. Numerous official attempts, with police and health authority backing, failed to cut down on the overcrowding in a dark chamber devoid of canteen or any other facilities, beyond two 'closets' with sackcloth doors.

Observers agreed then (and re-examination seems to support them now) in seeing, under Bristol's public façade, much depression – 'more than in any other area studied in recent months'. 'Open defeatism', in general very rare that spring, was encountered among younger workers; and the then Lord Mayor mentioned this personally to one of us. There was a notably keen following for the nightly broadcasts from Germany by 'Lord Haw Haw', the Irish-American William Joyce. At the same time, the Regional Information Office in Bristol reported deep disillusion with BBC news bulletins, largely and as usual because these were regarded as doing insufficient credit to Bristol's sufferings.

A sense of insecurity grew from the earlier complacency. Grumbling against authority became unusually noticeable. Before the last big attacks, trekking and personal evacuation had both become common, while among the stayers into spring a familiar mood was monitored: 'People are getting worn out; the irregular, sporadic, sudden switching on of heavy raids here has a strongly disturbing effect. Frequent attacks are not a help unless they are regular or come on towns of over half a million inhabitants. While it would probably be possible to condition people if they were given help and leadership, it is not an automatic process as some people seem to suppose.' Insecurity was accentuated when the 162-plane raid of 16 March was interrupted by an all-clear. 'Many let out a great sigh of relief' and went from shelter to home or basement to bed, as the attack reopened with violence.

Ten days after the last big raid, in April, damage remained 'the dominant topic', approximately 50% of all conversations overheard, mainly personal accounts of dangerous experiences. This was an extreme prolongation of the phase. There was also spontaneous comment on the inadequacy of shelters, the need for more tunnels. These tunnels were being improved, bricked and cemented by now, though to do it some shelterers had to be moved out temporarily – again a source of fiercely unfavourable talk.

Remarkable on the positive side was a considerable effort by the Regional Information Office to inform and actually to amuse the blitzed stayers. Meetings and concerts thus organized 'were received very favourably'. Bristol night-life, exceptionally, 'improved' in late winter, with cinemas 'nearly normal' through February, though they had been in bad shape before Christmas. Matters came to such a pass that the Lord's Day Observance Society, seeing serious encroachment on their reserves, held meetings at which they threatened dire judgement from above if sinful sabbath-breaking persisted – with a distinct hint that the Luftwaffe's return might result from Our Saviour's displeasure.[4]

That this Luftwaffe was no respecter of Sabbath was for sure, as Sotonians had learned. 16 March was Bristol's Sunday night of suffering, too. The last two raids fell on week-nights. By 12 April, a Saturday, the worst was nearly over. But across the south-westerly prolongation of England, back on the south coast, Plymouth, with not much over half of Bristol's population, was still due to receive five major raids after that, from over 600 bombers, once doubled, once trebled, all on week-nights. Nearly everything that happened in the south before, rehappened in Plymouth. We cannot do better than move there to conclude this part of the discussion: It is a lively story . . .

(5) Plymouth – Drake's country (& Astor's)

'What is the spirit of Drake which still persists in Plymouth, and the spirit of the Navy? It is a strange thing, invisible, intangible, imponderable, and as we know now, of an audacity well nigh incredible.'

LORD ASTOR, 1945[1]

'Even Sir Francis Drake and Britannia are sticking it up on the Hoe.'

'STAINLESS STEPHEN',
16 January 1941

In these days of massive 'social science' little attention has, so far, been paid to the measurement of those tangled traditions, experiences, sentiments, relationships, emotive fantasies and ecological facts which determine local sentiment and the parameters of local loyalty, as compared with wider attachments such as 'race' or country. In 1940 you may believe that the void was total. So one can only generalize vaguely, with no more than intuitive support, when suggesting that a feature of Plymouth in the thirties and forties was a particular form of provincial, self-conscious pride. Coventry had its cathedral, Portsmouth its *Victory*, Southampton and Bristol nothing so tangible as a centralizing symbol. Plymouth somehow had something more. Difficult though this may be to assess, the point cannot be ignored in judging how people lived through their blitzes.

Quite apart from any such subtle sentiment, Plymouth's blitz pattern was different from the other southern ports – and most of the country – only one 100-bomber raid took place in November 1940, then seven in March–April. The main Mass-Observation team came on from Portsmouth direct to Plymouth in January to find 'cheerfulness everywhere and a real spirit of "carrying on"'. 'Not that it is fair' (we noted) 'to compare Portsmouth and Plymouth directly'; in the latter the damage was as yet far less.[2]

The real strain came with the start of spring. During the daylight of 20 March H M King George VI and his Queen visited Plymouth and took tea with Lady Astor on the Hoe after touring the town where they had been 'well received'. They left at 6 p.m. That night and the next the place received its heaviest raids, 293 planes, 346 tons. Our File Report 626 takes up from there, following a framework liked by our clients of 1941:

The present report is based on investigation from immediately after the second raid until Sunday March 30th, undertaken by four investigators, three of whom were responsible for the previous report on Plymouth.

Plymouth morale is still, on the whole, good, and in our estimation ranks with Liverpool, when allowance is made for the very serious damage done to Plymouth. At the same time, there is an appreciably large and qualitatively important group which may be summarized:

a. Good morale, especially among those who have stayed put in the badly hit areas, and the men, particularly men in pubs, etc.
b. Moderate morale, among the middle-classes and some of the older people.
c. Definitely bad morale, conspicuous among evacuees and among women who have stayed behind. (And so on . . .).

Next came the five nights of a late April week, an intensity as yet unprecedented in a smaller city. FR 683 of 4 May 1941 reported the human effects with due weight and an anxiety suitable to the occasion:

The Plymouth situation is, from a long term point of view, the most serious yet. It opens up a new vista of civilian war. It possibly indicates a new policy for enemy air attacks; the incessant blitzing of a town for nights on end, then a lull of weeks, then another series. As we have pointed out in earlier reports, this is much the hardest type of attack to get used to. Many people appear to find it impossible to condition themselves to this rhythm. *And in any town of much under a million inhabitants, it is very doubtful if people can condition themselves to more than three nights of intensive attack in a week.* That is to say, people cannot condition themselves if it is *left to themselves* to do the conditioning as it is now. If they are left passive, relatively defenceless, and without any alternative outlets from the emotions and repressions which inevitably follow upon nights of random death and a destruction which is far more amazing and terrible to the ordinary people who have lived all their lives in the town and perhaps in the same street – then they have no answer to their local disaster. They are not defeat*ist* – far from it. But they are defeat*ed*. Their town has suffered a major military defeat; and they, the untrained and undisciplined soldiers of the Home Front can (at present) do nothing much about it.

Compare the diary of Harold Nicolson, at this time a member of the Civil Defence Committee of the War Cabinet, written (with an eye to publication) three days later than the above extract, when he notes that Herbert Morrison is worried about the effects of provincial raids on morale and 'keeps on underlining the fact that people cannot stand this intensive bombing indefinitely'. Morrison concluded that 'sooner or later the morale of other towns *will go, even as Plymouth's has*'.[3] It had taken the minister nearly six months to recognize openly a situation already spelled out from Coventry on. Now he went to the other extreme – and sent a whole city's 'morale' likewise. Did 'it' all of a sudden go to hell?

By now, Germany's aim was growing as clear as it ever would be. In twelve weeks from 19 February to 12 May, 39 of 61 raids involving over 50 aircraft were directed against the western and south-western ports, from Plymouth through South Wales to Merseyside and Clydeside. London received only seven raids on this scale during the same period. The emphasis was on the trans-Atlantic life-lines which the U-boats had not been able to cut.

Only the week before Morrison's personal panic, his own chief contrastingly declared on this struggle, over the radio, that those men, women and children who had suffered the most severe bombing shared the 'most high and splendid' morale. This concept caused Churchill to feel 'encompassed by an exaltation of spirit in the people which seemed to lift mankind . . . into the joyous serenity we think belongs to a better world than this'. His sublime vision was richly coloured by a visit to Plymouth, where a thin crowd expressed themselves somewhat differently.[4]

Before seeking to balance, if not resolve, the extremes of serene (Churchillian) and moraleless (Morrisonian) assessment, that visit is worth swiftly following. The Prime Minister's tour, due to start at 3.40, began at 4 p.m. on 2 May. Just ahead of time, police with loud-speaker vans announced he was coming. By then, thousands of afternoon trekkers were already on their way to night billets, so that only a small proportion of locals remained to see him. About 500 did so at close range, 3500 more distantly, on his extensive but rapid trip. The open car moved fast, stopping twice in the hour, with Winston perched on the back in Trinity House uniform, Mrs Churchill alongside and Lady Astor, Lady Mayoress and MP (Lord Astor was unwell). Cheers, though general, were scattered and feeble. Outside the Guildhall, however, about 120 showed 'a deeper and more moving' response.

'For those who really got a see and sniff of him, the experience was profoundly moving.' He brought (we thought) a breath of positivity, vitality and above all *concern* into what appeared by then to be 'an atmosphere of negative devastation'. As we then saw it:

> It was impossible even for the most objective observer not to be deeply moved by the sight of this great man, fierce faced, firmly balanced on the back of the car, with great tears of angry sorrow in his eyes. He was so visibly moved by the suffering that he saw, yet so visibly determined to see that it spelt not defeat but victory. A man overflowing with human sympathy, an epic figure riding through this epic of destruction.

The Prime Minister was overheard to make two remarks:

'God bless you all' (with tears in his eyes).
'Well done, Plymouth' (ditto).

Onlookers responded with: 'Good old Winnie' or 'He's like a *lighthouse*.' There was much pleased feminine comment on Mrs Churchill's striking appearance: snow leopard coat with flounced sleeves, a coloured handkerchief over her head with phrases from her husband's speeches printed on it. Some were surprised Winnie had such a fine, fresh-looking wife.

A smaller number of onlookers and many more who heard of the tour afterwards were influenced by the unhappy aftermath of the royal visit in March. 'We'll have a raid tonight' was a common sentiment. But there was no raid. Plymouth's nightmare was nearly over, after a hard course; 1172 dead.

Many had convinced themselves through that quiet February that nothing worse would (or could) come. In that month Plymouth was gay, night-life near normal, the pubs full of song – with 'Bless 'em all; the long and the short and the tall' most popular. At a time when Portsmouth or Southampton were ghost cities, Plymouth was pulsating. The Palace, a classic variety hall, played to full houses. The show 'Go to It' was a great success, featuring comic 'Stainless Stephen', a sort of sad-sack cynic forgotten now but so important in the war that, with Tommy Handley of the BBC 'Itma' series, he rates entry in the highly selective *Biographical Dictionary of World War II* by Christopher Tunney, 1972. Born at Sheffield (thus the 'Stainless') in 1891, he wrote his own material in his own special dialect, including here many topical or local Plymouth jokes adapted to the previous raids:

'I've only had one hot meal since Sunday – horse-radish and a rissole.'
'What does it matter if Germany does invade us. They won't get across Union Street' (for the damage).

That evening's biggest laugh was at one of several transport jokes describing the sluggish Plymouth trams as 'those instruments of death that flash past'. He flattered the courage of

Plymouth people; for instance, their knowledge of the value of money: 'Twelve pennies make a shilling – twenty shillings make a pound – £5000 makes a Spitfire – the Spitfire makes thirty Messerschmidts' (a big laugh). Stainless told of visiting Devonport dockyard, to find all the flags out and drums beating; he asked a sailor if there had been a famous victory:

'No, two packets of 20 Players arrived in town' (huge laughter). Plus touches of what counted as pornography in those pre-permissive days – such as the joke (cited above) about Drake and Britannia sticking it up on the Hoe (for Britannia read Nancy Astor in local code).

Self-congratulation, comic or serious, was much the order of those intermediate days. In mid-January the *Western Morning News* proclaimed:

'PLYMOUTH STOOD THE TEST'

– supporting this with a congratulatory statement from the Lord Mayor, Lord Astor. The newspaper's special reporter (and later local war historian) H. P. Twyford, put it another way after the unfortunately-timed royal visit, when a sassy warden spoke to the Queen a few hours before the really big raids began:

A R P Warden: 'We are keeping our chins up.'
H M the Queen: 'Well done. It's only by keeping one's chin up, as we are doing, that we shall win the war.'

But 'at 8.30 that evening', Twyford notes, 'the inferno was let loose ... Plymouth's heart was torn out and mauled'. Her people (he wrote) saw the 21 March dawn 'with hatred raging in their hearts against the vile perpetrators'.

Sadly, suddenly, these people learned that what had gone before was by no means the worst that could come, as they had lulled themselves into believing. Nevertheless, and once again, the immediate response was to feel they'd had their show. 'We've had ours' and 'after what we've been through' were common phrases of late March. Yet others were uneasy, especially that not too much damage had been done to Devonport docks ('so that another raid is likely'). The March damage was visibly concentrated now, in the heart of the city, previously spared. A large proportion of homes were not yet too seriously affected, however.

As usual much went wrong with the post-blitz social services. The savage March double assault caught everyone by surprise. More realistic reactions now began to pervade the city, with its whole centre cordoned off for days for all to see. Overheard comments from working men commonly went this way:

'We never kept up our fire precautions as we ought to have. That was the trouble.'
'It serves us right in a way for getting so behind.'
'We here were getting too, what do they call it, complass-ent. We'd got so that none of us expected any raids. Then we had a rude awakening.'
'Well, what can you expect. It's our own fault for being slack.'

This extensive self-awareness, seldom encountered elsewhere and never on this scale, arose out of the exceptionally light attacks through the winter as compared with similar ports. These had been mistaken in Plymouth for more, quantitatively, than they were. A lot of people now realized they had deceived themselves. With this awareness went considerable pessimism about the future. The official casualty list, with 336 killed during the month, covered twelve pages posted up publicly, though these official figures were seldom accepted: the dead were frequently said to number thousands. 'Thousands. It's so depressing to think of it.'

Raid talk regained its obsessive quality in the last week of March, rating up to 80% of all talk overheard six days after the 20–21 March blitzing:

1. 'I see a Churchill dead in the paper this morning. Result of enemy action.'
   'It's the fishmonger.'
   'I thought he'd be about that age. Poor old Churchill.'
2. 'I wasn't in a shelter either. Never have been a shelterer . . .I don't mean I was outside all the time, mind. I was in a house near the park a bit of the time. And I had a drink – a couple of Guinnesses – when it was all over.'
3. 'I can hardly *recognize* anything. Lots of things you can hardly recognize.'
4. 'The whole place was aglow. You could see to read a newspaper.'

The idea that Plymouth was the worst hit place in Britain now gained fresh (and more justified) vigour. 'The way in which the people wallow in their own destruction' had already been conspicuous back in January, when it was assessed (and stressed) as 'likely to end, unless attended to, in a severe reaction if a much more destructive raid took place'. Nothing of that kind was attended to; and much more destructive raids did take place. Now the process was to repeat itself, stepped up – rendering Plymouth more vulnerable to the most severe raids still to come in April.

Though a conspicuous number were, by April Fools' Day, trekking or quitting, sleeping even on the fringes of Dartmoor, much of Plymouth stayed put. Ten out of fourteen houses in one heavily bombed working-class street had their washing up in their backyards on Wednesday, the main drying day in local tradition (on the conspicuously high lines they then used in those parts). In another street, seven out of sixteen had the laundry up. But near the centre of town, on the fringes of the burned and blasted zone, only one or two in ten showed laundry. One shop, its owner trekking, left it up to the neighbourhood at night:

IN CASE OF FIRE
PLEASE GET DRUM OF PARAFFIN OUT
IT IS INSIDE WINDOW
ALSO CAT

Watching the damage took some fresh forms in Plymouth, too. On one corner, a mantelpiece with a clock still going stood stark upon a shorn-off wall. The clock interested practically every passer-by during several days. 'Over and over again' it was the subject of 'delighted comment'. Simple freak views of this kind gave a relief near to Stainless Stephen's irreverent comedy.

More elaborately, the RAF, taking one more rare post-blitz initiative, brought in a German Junkers 88 dive-bomber, exhibited unlabelled, on 27 March, in the more damaged part of the town. Crowds examined this, from 2 to 20 minutes each person. The overwhelming effect was of mild pleasure, interest and relief: relief at something which could be seen and understood, the lighter touch which made (as one observed) 'the vast chaos all around somehow more real, intelligible and concrete to the still amazed inhabitants of Plymouth'. Men were more interested in the material, engine, craftsmanship, all of which were repeatedly praised. Women noticed especially the size: 'many seemed amazed at the size of the plane, the length of it, the weight of it.' Some evidently hitherto had thought of an enemy bomber only as a remote thing, a spectre in the sky, a moth searchlight-beamed. The reality was somehow reassuring, almost friendly. 'In hundreds of overheards collected; none were hostile and none inferred reprisals.'

The displayed Junkers showed no clear signs of damage and was generally considered to have crashed. This led to quite strongly voiced comment on the inadequacy of anti-aircraft defences, now widely criticized along with surface shelters (which were often water-logged and doorless) and the shortage of most foods in town ('even a packet of chips'). Stainless Stephen's joke about the packet of fags had grown less funny. Once more there had been a marked absence of canteens – mobile or fixed – except for those reserved for the armed services. As the gas supply was cut almost all over town, cooked food became a major lack in the last week of March, just as it had before in Portsmouth, Southampton and Coventry. It seemed hard to believe that those early, easy blitz-lessons had not been learned, still less applied.

Local limitations were no longer unexpected by those few of us who had followed these sequences of sadness as they came. But nearly seven months after 7 September 1940 the low threshold of reaction in London itself was not so readily intelligible. Did everyone have to learn everything each time afresh?

The Ministry of Health, responsible for a wide spectrum of after-raid welfare, was particularly slow in recognizing and then meeting at the local level the need for its services both to exist and to be seen to exist in situations like Plymouth's. Their first circular (No. 2269) to provincial authorities on the need for information and administration centres went out from Whitehall in the later half of January 1941. Their first circular suggesting, in parallel, the need to appoint special welfare organizers was issued on 11 August 1941 (No. 2453). The needs were known in November 1940, at latest.

Credit is due to the Ministry of Health for their double-crown posters listing Rest Centres and places to go for specific services and information. These appeared all over the place on 29 March, sometimes alongside notices closing down *all* places of entertainment regardless.

PLYMOUTH CITY POLICE

I hereby prohibit the use of the premises known
as . . . . for any entertainment
to which the public are admitted, until
further notice.

G. S. Lowrie
Chief Constable

Shades of Stainless! This left nothing sociable to do of an evening but drink, though, if anything, *less* drunkenness 'was observed in March than in January'.

The absence of night-life, as in Southampton – but here deliberately imposed (for no clear reason) – cast a deepened gloom over the city. Similar restrictions, intentional or unavoidable, limited other personal outlets. For one, the local newspaper continued publication through March, but printed later on in Tavistock out on the moors, where the Deputy Regional Commissioner, sent from Bristol HQ, and the senior officer for Health also set up shop. Some evenings not a single paper-seller was seen. An observer who had just obtained back issues of this paper went down the street with them, to be besieged by people wishing to buy copies, though it would be difficult to imagine anyone less like a newsvendor than this conservatively dressed 6 ft 4 in Old Etonian.

Lack of information, lack of news, uncertainty, anxiety, breed abundant rumour as relief. The commonest rumour in Plymouth referred back to an alleged broadcast on the unfortunate royal visit of 20 March. These reflected bafflement at trying to understand the Luftwaffe system of target selection and timing. Many who had already decided they had 'had their fair share' groped for a rationale to the raiders' return after such a long lull. 'Broadcasts' announcing the royal visit were blamed in arguments lasting all the following week. Some blamed the BBC, others the magic insight of Lord Haw Haw. The general view was that rash publicity of some sort had brought this swift reprisal. The origin of the story probably came from announcements by Radiodiffusion, a then young internal switch-system run by Broadcast Relay Services, which was often able to keep going when the BBC had to close down. At one stage when the local electricity failed in January this firm had been supplying its own generated current to the Town Clerk and Chief Constable. Although a plan to link up officially with Radiodiffusion in Plymouth was apparently rejected on policy grounds by the Ministry of Information's Bloomsbury headquarters, it developed a special relationship with the local authority, and eventually made an important contribution by carrying ARP instructions, five-minute talks by Lady Astor and others, over its commercial network – and in violation of its licence conditions.

The Town Clerk of Plymouth, like his peers in other towns, enjoyed great powers of local decision-making. He was evidently among the most able officers of his class in blitzed Britain, with strength of character and quickness of mind. Called 'Pooh-Bah' like his colleagues elsewhere, much was left to him by his Lord Mayor, Lord Astor, a gentle man with many wider obligations. Some said he was left too much to his own devices. Moreover, the Emergency Committee set up to deal with crises was strongly criticized in informed circles. One of its three members was over 80; only one was regarded as a 'really active figure'. This was seen to facilitate 'the complacent dictatorship of the Town Clerk'.

The Lord Mayor's Tory MP wife (the first woman elected to the House of Commons) was herself widely spoken of as an activist: 'in the forefront of blitz work'. Some comment was mildly malicious, but in general people were 'truly admiring of her personal energy and genuine interest in the people'.

Whether or not things were really so lopsided at the top, there was wide unease that the powers that were 'had become too complacent and were inadequately prepared' if worse were to come. It came with the five very harsh April nights one month after the royal visit and subsequent raid.

On 22 April 1941 Twyford wrote that: 'Last night commenced another series of appalling raids and until the end of the month the city was to pass through yet another terrible ordeal which was to test the spirit, the endurance and the courage of its citizens to the limit.' By month's end most Plymouth houses were in some way damaged, 8% totally destroyed and 16% more uninhabitable for at least two years. One in four was thus rendered homeless in the technical sense. After the last of the five big April raids Mass-Observation found a great contrast with its previous Plymouth report: 'The civil and domestic devastation exceeds anything seen elsewhere, both as regards concentration throughout the heart of the town, and as regards the random shattering of houses all over. The dislocation of everyday life also exceeds anything seen elsewhere, and an enormous burden is being placed on the spirits of the people.' Walking the littered streets on 30 April, a Wednesday, there was no one about except sailors or youths. A group of highly made-up girls were swinging down the once-main shopping street singing a recent jazz hit which might have been designed for the occasion:

'We'll dine at the old café,
Where we had so much fun...
We'll order cocktails for two.'

The punch-line came with the chorus's end, roared into a dull sunset:

'Painting the town the way we used to do...
For I'm stepping out with a memory tonight.
For I'm stepping out with a memory tonight.'

Girls like these behaved wildly, accosted boys, showed, by the standards of the time, 'signs of considerable moral abandon'. It was a feature in the last stages at Plymouth, and something rarely seen in 1941.

Those were the gayer ones! The general evening atmosphere was intense, deep gloom, charged with the expectation of perhaps a sixth night under fire. Now came 'the occasional murmur of hurrying footsteps', through once busy, crowded places. It was a positive relief to one observer when a passing girl asked her companion:

'Ain't it comin' on windy, eh?'

Round the corner he quickly came back to the familiar blitz-line:

'I never expected to come out of it alive.'

And then, going into a pub, came what really could have been Plymouth's heart-cry:

M.1 'Where's the gents?'
M.2 '*Everywhere's* the gents now.'

Pubs remained, as before, 'oases in chaos'. Earlier concerts and meetings had disappeared, along with the rest. There was precious little laundry out along the lines that Wednesday.

Back in late March it had seemed necessary to distinguish three identifiable groups in their reactions to developing events. By the end of April these had budded into five more or less distinct categories, though complex and overlapping:

(i) A 'stay put' population, staying *voluntarily*, and refusing to budge from Plymouth whatever happens. This is the most conspicuous group within Plymouth, because it carries on its activities as normally as possible, and makes up practically all the people who now walk the town at night for pleasure or duty. A large part of the gayest and bravest of these are young people, and we especially noticed the high spirit of young girls (aged 14–22).
(ii) A resident population which would like to get out, but can find nowhere to go, cannot afford, or has to stay for other reasons.
(iii) The nightly trekkers, whose homes are more or less damaged and unsuitable for sleeping in, i.e. compulsory trekkers.
(iv) The nightly trekkers whose homes are undamaged and who go out because they are 'afraid' to stay at night.
(v) People who have left the town altogether. Many of these have their homes not seriously damaged, and they include appreciable proportions of air-raid wardens, fire-watchers and others with responsibilities, though on the whole these people with jobs to do in a blitz have the best morale of all, precisely because they have got something to do and do not wait passively.

Raid talk was more than ever preponderant: 'endless discussions of incidents, damage, terrible sights, marvellous escapes'.

Gas-mask carrying reached the highest level for any provincial town. Local boasting about courage and the like was barely discernible. Instead there was heavy emphasis on the suffering. The local press wallowed in it:

PLYMOUTH ASSUMES MANTLE OF COVENTRY

And:

CITY DEVASTATION WORSE THAN COVENTRY – SAYS US ENVOY

Or again, Plymouth:

'ASSUMED MANTLE OF COVENTRY': LONDON OPINIONS

We were able to spend an afternoon with eleven national and local journalists (including historian Twyford). All but one held to the opinion that Plymouth had been very much affected by the April raids. They were unanimous in expressing dissatisfaction with measures to meet the situation. What they said was, of course, unprintable in the press in wartime.

One of us also overheard Lady Astor reproving some of these reporters for the emphasis that had been put on what she called 'little things' – like the continuing refusal of private motorists to give people lifts out of town. Though she chided them in a bantering way, several of them were furious. She separately made an outburst against the 'Ministry of Inflammation'; because of the way it *under*-played the attacks in the B B C news, etc.

Another observer spent a morning at this stage with two senior officials concerned with post-blitz problems. They both thought Plymouth's will to carry on had been seriously weakened, although the people themselves remained (they thought) 'uncomplaining, grateful for everything done for them, determined to make the best of things'.

Officials of four Rest Centres, a leading Nonconformist, an Anglican priest and a police officer with fifteen years' local experience echoed these views. Two top business men agreed: 'we're all scared stiff' and 'there's no city *less* prepared than this one'. The policeman was practically defeatist: 'people have just had enough,' he said. He considered open 'defeatism' possible if there were more big raids. But there were no more.

Nearly everybody by now had changed tune and expected more to come. 'They'll let us get cleared up first, before they start again'; or 'as soon as we've got going, they'll be back again', were repeated echoes of spring's deep anxiety, near hopelessness. Many firmly expected the 28–29 April attacks to go on on the 30th, when, by 10 p.m., tension could be most clearly felt 'even among those who so certainly and gayly stayed in town'.

Nothing was felt to be secure. There looked to be no real answer to future 'nights of hell' – this last the noun most often used. Indeed, the five late April raids had lasted over 23 hours, with some 1000 big bombs, 17 landmines, thousands of incendiaries, just under 600 dead, rather less hospitalized.

Similarly, there seemed nowhere really safe and within work-range to 'escape' to. Surrounding billets were full to bursting. Previous lack of confidence in shelters was acute. Among those who stayed firmly put, there was argument on the pros and cons of remaining indoors or in shelters; but with little enthusiasm for the many brick surface shelters now available. These comments will illustrate the range:

Worker 25: 'I'm not so keen on shelters. A hell of a lot of shelters have been hit. I've heard 102.'

Worker 45: 'I always go to the one under the railway bridge. I reckon it's as safe as any. Of course it doesn't matter where you are if you get a direct hit, but there's no sense in staying in the house if you can go to a place stronger – no sense in asking for trouble.'

Woman 65: 'Well, I did apply for a shelter, but there wasn't room at the back, so I just sit in the kitchen. I daresay I'm as safe in my own home as I am in a shelter. But it makes my stomach go cold. I sat on the stairs paralysed on Tuesday night. I always have a drop of brandy when it's bad – with hot water. My doctor told me I might just as well drink water alone as cold water with brandy.'

It must be remembered at this point that most of the population in a place like Plymouth then left school at or before fourteen. There was no TV, foreign travel was a middle-class luxury, a week by the British seaside the normal annual liberation. The 'ordinary' Briton, housewife particularly, lived in the shadow of a complex of insecurities, of health and money, unmothered by any welfare state – yet in the certainty that whatever happened she could get up in the morning, go down and turn on the tap for water, find gas to boil it, walk round the corner for the Rinso or that packet of fags. In 1941 this pattern vanished, almost, it seemed, beyond the point of no return. But the less educated held on strongly to a basic optimism beyond which it could be fatal to go. Approaching that point of breakdown was this overheard voice, telling another (1 May): 'Terrible. You can't pick up the *threads* of your life.' A naval Petty Officer to his landlady:

'I can't stand this much longer. I'll have to go away.' (1 May)
'I'm going down to Cornwall for two days. I'll have to get out of this. I can't stand it. It's affecting my kidneys already; my water's like mud. And once your kidneys have gone, you're finished.' (2 May)

And he went...

Whether down to Land's End or up to Dartmoor, getting away from chaos became the finally logical answer, when not only did practically nothing function normally at home but little seemed to be being done to make things function soon. After all, as Anthony Powell remarked in his novel *Books do Furnish a Room*: 'It is not what happens to people that is significant, but what they think happens to them.' Extend that to 'and what they think is GOING to happen to them' and you have the crux of living with, through and out of a serial blitz like Plymouth's that April.

It is in a sense astounding that anyone, let alone thousands, stuck in Plymouth, was able to carry on so capably in such a physical mess. The stayers were normally, audibly, proud to be there. They expressed contempt for the rest, especially quitting sailors ('yellow'). But those who left often did so simply because they could see no other way to continue. Due largely to the official refusal to reach any public decision on evacuation, the muddle was immense. We found late in April: 'The chaos of evacuation arrangement ... was so obvious that it even got into the (censored) press. *No* ordinary person in Plymouth can understand why until the last severe raid was over it remained A NEUTRAL EVACUATION AREA.' The same documentation went on, after seventeen close pages of observation, to apologize that now 'personal feelings' would intrude a little: so bad did things seem even to the marginally involved. FR 683 of 4 May 1941, after emphasizing our rule of attempted objectivity, went on:

For once we beg permission to break this rule, and indeed it is desirable that we should do so and admit the possibility of our own emotional bias in this matter. Let us, therefore, frankly say that we regard the whole past story and present situation of evacuation in Plymouth as disgusting, degrading and tantamount to sabotage of the war effort for those responsible. Who can let things get into such a mess, and get away with it? The Minister of Health was in Plymouth himself on March 24th, and according to *The Times*: 'The question (of making Plymouth an evacuation area) was then discussed on the spot, and it was agreed that it was not necessary or desirable to take action.' There is little sign that the local authorities have been making any violent fuss about the matter since then, though there is now a tendency to put all the blame on the Ministry of Health. The real explanation perhaps lies deeper. Perhaps it lies in the absence of any high policy from above, any high strategy of the civilian war which gives a background for decisions, a yard-stick for the considerations and priorities of human welfare on the Home Front. Certain it is that the matter has got completely out of hand and among people who are getting blamed now is the Regional Commissioner. It is reported that not even the Registration forms for evacuation were ready in advance.

The only 'proper' way for a homeless person to leave blitzed Plymouth was illustrated by a typical case-history (at page 26 of that same document):

Take the case of a sailor's wife, her husband at sea, she keeping a small shop, which is gradually dwindling as all small shopkeepers are frozen out by new Government orders. She has two children of school age, and a baby. The shop and the rooms she lives in above it, received a direct hit. One of the children is killed, another slightly injured. She is naturally extremely shaken. She is living in a badly hit working-class area. She finds herself homeless, moneyless, without a thing except her spirit, her morale, her common sense. What does she have to do next day?

(i) She has to eat. She has to get new ration books. She goes to St Bartholomew's Hall, Milehouse.
(ii) In the meanwhile, she has to find a communal feeding centre, for the children, go to the Girls' High School.
(iii) She has to try and have tinned food and other stuff in her small shop in ruins salvaged. All right, she has to go to the city's Treasurer's Office, whose last address given in an MOI leaflet is the Guildhall, but that got bombed on the previous night.
(iv) She is also supposed to report the presence of some contaminated

food. Here she should contact the Medical Officer of Health, Beaumont House, Beaumont Road.

(v) She is already beginning to be faced with the problem of her surviving children as she treks around. She has to find a Rest Centre – the ones in her area are destroyed. She finally leaves them there for the time being.

(vi) But one child is dead. She has to locate the body. She has to inquire about the death certificate then, at the Information Bureau in the Technical College.

(vii) She decides she can't afford a private burial, and so applies to the war deaths' department in the Information Bureau.

(viii) Now she is faced with the problem of extra clothing especially for the children. The address is 18, Addison Road, Sherwell.

(ix) She is advised to put in a claim for compensation for clothing, furniture, etc. So off to Mutley Baptist Church, the Assistance Board.

(x) They are able to give her a cash advance, but necessarily limited. They advise her to apply for extra help from the Mayor's relief fund, situated in Morley Chambers.

(xi) As it is obviously hopeless to get any alternative accommodation locally, she desires to apply for free travel voucher for the homeless. The address is 'Seven Trees', Lipson Road. Or a child evacuation form can be obtained from the Education Offices, Cobourg Street.

(xii) Having lost the identity cards, naturally, she has to apply for new ones to 13, Thorn Park, Mannamead.

(xiii) Then to get new gasmasks. Off she goes to the Corporation Surplus Stores, Mill Street.

(xiv) Of course, she has lost her 'Navy book', and to get another she goes to the Pennycomequick Sorting Office in Central Park Avenue. Having got all these Plymouth matters in hand, *the problem of her billet in the country arises* – a terrible problem for the good citizen who has stayed put until actually bombed out. Here the Chief Billeting Officer, Lipson Road (no address given on MOI leaflet) appears to be involved.

(xv) Finally, being a good citizen, she decides to avail herself of the advised post-blitz measure of typhoid inoculation. The good citizen goes to the Prince of Wales Hospital, Green Bank.

But what a good citizen this would have to be! ... In practice, at least three in four quit long before that finish, casting themselves upon the countryside. In this, they were not much helped by Plymouth's hinterland, with few suburbs or satellites, a great deal of wild moorland sparsely inhabited, villages small and far between. To have provided full reception facilities for the homeless alone would have taxed the rural background heavily. 'Some distress was therefore inevitable', as Titmuss remarks for Plymouth. But, as he goes on to admit, distress was compounded by a failure to prepare properly, mostly due to 'lack of co-operation between authorities for town and country areas'. In some cases it was not just lack of cooperation but positive non-cooperation, bordering on hostility.

Once a citizen had been fortunate and clever enough to find somewhere to sleep after the earlier raids, she or he might well share the experience of one local observer who visited many outlying villages:

In the 'rest centre' at the church hall there were, at 10 p.m., 21 women and about 15 children. There were also five men and a boy of about 15. This was a one-room rest-centre. At the top there was a platform on which about half a dozen women were preparing sandwiches or pouring out tea or just waiting to hand people what they asked for. Sandwiches, tea and buns were set out on their trestle table.

The parson (about 45 years of age) came on the scene at 10.15, holding a couple of soup plates. Said he: 'Just a minute everyone please. You see I've got here two plates of soup, nearly full. And bread. Well, someone's left it. I'm not blaming anyone here – the person responsible may have gone out. But I do appeal to you not to take anything if you don't want it. There are plenty of hungry people who would like and who need food – I wouldn't mind this myself. But to leave food is just damned selfish – excuse the expression. But it's very disheartening. All the people here are doing their best to make you comfortable, and if you'll just co-operate, they'll be very grateful. Thank you.'

No beds, no bunks; nothing at all to sleep on but benches and the floor. No mattresses. The women were sitting around mostly, but a few had put their own material on the floor and were lying on it covered with blankets supplied by the centre. All the men were sitting around. I soon knew why – why they weren't lying down. For about 10.30 the parson reappeared: '*Will all the men leave now, please.* The men's centre is at the Anglican Church up the road on the left. They can get blankets here.'

The observer followed up the dark path:

Four men went over and got blankets (two) followed by M, and a boy of 15, presumably his son. I followed them up to the Anglican Church. A local fellow standing around opened the awkward little gate, and led the way into the Church, 'Here you are,' he said, 'Anywhere you like.' And he cleared off, banging the door behind him. I was amazed. Here was the 'Men's Rest Centre'. A pitch-black church – not a glimmer of light anywhere. A cold church. A church equipped with narrow pews – that are hard enough to sit on, and impossible to sleep on. The man and boy didn't stay long. No sooner had the local slammed the door but the man declared, 'I'm not staying in this bloody place. Come on son, we're getting out.' Another man followed, saying, 'I'm not sleeping in a bloody church, either.' The rest presumably lay down – I couldn't see them, but I heard.

No wonder then, that only a tiny fraction of these urban refugees so swallowed their pride as to stay in whatever public accommodation was supplied rurally – and we tried out several more, including Tavistock and Roborough Village Hall, equally miserable. Instead, thousands preferred to camp out, sleeping wherever they could get to, including on Dartmoor.

Another observer joined a group of some forty refugees on the edge of moorland:

They were walking about, and I asked them what they were going to do for sleep. 'Oh, we will be all right. The moor soon dries off, then we will roll ourselves up and keep warm.' All the time I heard only one complain, why were the people not ready to get us away after the first big raid. Those who had been bombed had a lot of praise for the soldiers in the way they helped and moved their furniture and things to other places for store. The general talk was still about the Blitz and all the details of who they knew had been killed and how the incendiaries had been put out. Some began to cuddle their children up in rugs, others still walked about and quite a few started *back to town.* As it began to get bitter cold about 11 p.m., and by that time it had dried off, I sat down with a few men and women who were smoking. The children had been rolled in the bedding. This group was bombed out and were getting their children away in the morning, but it was hard work getting them to talk, they just sat and smoked. I tried to snatch a sleep when the men laid down but was too cold so went walking. After a time others got up and by daylight there was a general walk back. The children had some sleep but not many of the grown-ups did. Most of them on the way back said they would not come again after the children had gone. I don't know how they kept cheerful because it was frosty and I was dirty, tired and frozen.

Those, fewer still, who chose – or could not avoid – to use such Rest Centres as existed back in town, were likely to be as uncomfortable. 'Perhaps the most moving sight we have seen in all our investigations' (we wrote on 3 May) 'was eight old women, all over 70 and all dressed in Victorian bonnets and lace, sitting in a large bare underground room which is one of Plymouth's (official) rest centres' for the homeless. They had been on the hard seats (there were no others) for five days. Their 'home' had received a direct hit, killing some, injuring many, leaving nine intact. No billeting officer had been. Nothing was yet arranged for their movement. One survivor had 'gone off her head'. She had been forcibly removed from the presence of the rest.

From the fire, debris, dereliction and distress in the south emerged outstanding exceptions where order was captured from the edges of chaos: the RAF *coup* at Southampton, the Citizens Advice Bureaux in Portsmouth, and at Plymouth, strikingly, the Assistance Board. At first the Town Clerk there had characteristically included responsibility for this within his own Pooh-Bah Kingdom. The Assistance Board, a semi-autonomous body closely linked with the Ministry of Pensions in London, sent down a special officer to take over as things deteriorated. He was given authority and a mandate to act in ways he thought best for emergency solutions. This intelligent, warm-hearted individual provided leadership and decision coupled with experienced skill. Yet he and his staff did no more than their duty, mellowed by common sense and a certain charity in the best sense of that dubious word – made dubious largely by the track record of the Board itself and its predecessors.

Remember that in Britain at the start of the Second World War some 2,000,000 families were living from hand to mouth or from payday to weekend, according to our own observations and as best expressed by John Hilton's Halley Stewart lectures of 1938. Half of these were continually in and out of debt; mostly in.

Professor Hilton's 17% of Britons classed as truly impoverished lived 'out of sorts, out of future, out of favour – out of everything but wretchedness and debt'. Two million families had no margin whatever to save money on. To such as these, living in cities, the after-blitz mess could be largely a matter of added emphasis. Misery was accentuated, but it was not new. That, perhaps, made it easier to endure?

Many more, over two-thirds of the islanders on this estimate, in the late thirties had 'next to nothing in hand from pay-day to pay-day'. At the most, they might have a capital of £200.[5]

That scale of poverty helped produce a readiness by millions to accept a standard of living literally unimaginable to the better off, the 'fortunate million' of these pre-war, pre-Beveridge-welfare days. The loss of warmth, hot food, even furniture, was not a universal novelty.

It was primarily to assist the air-raid victims among the 'truly impoverished' that the Assistance Board system was reorganized out of the narrower function of Public Assistance and the old Unemployment Assistance Board, through the Personal Injuries (Emergency Provisions) Act and other measures designed for war-time. The Assistance Board was expected to need to give immediate cash help to those temporarily distressed because of war damage or injury, as well as to evacuate women with their children when in temporary need while separated from their husbands. Fundamental in this thinking, of course, was the official anticipation of the form of attack to be expected. This, as we know, overestimated fatal casualties or total homelessness, while belittling less drastic damage with its multiple repercussions on the mind.

As early as September 1936 the then small and new Home Office A R P department, with Hodsoll not yet Inspector-General, had a report prepared with a different emphasis. This paper anticipated that as well as major home destruction and mortality, there could be serious psychological damage; and that even in the smaller number of houses 'blasted down' the occupants might 'quite reasonably escape with their lives and even without injury'. Alas for the people of places like Plymouth, this more accurate forecast was overlooked in favour of the concept of holocaust.

Traditional, peacetime attitudes came homelessly to roost at Plymouth in spring 1941. Then, the Assistance Board in particular discovered that its organization was too slow, too narrow for blitz conditions. But they were still burdened with earlier decisions, such as the one which left emergency cash aid to them but transferred 'relief in kind' to local responsibility. Relief in kind included a whole parabola of post-blitz facilities. The Rest Centre system, for one, remained 'primitive, comparable to that of London seven months ago' (as F R 683 observed), although emergency feeding – answerable to the Ministry of Food – was much more effective, thanks largely to the Queen's Messengers, mobile canteens which really moved. But high-level confusion continued. On 30 April an independent MP asked the Parliamentary Secretary to the Ministry of Food:

> Kenneth Lindsay: 'Does this (control of canteens) vary with each authority? Is it sometimes under the Director of Education or is there some special officer responsible?'
>
> Major Lloyd-George: 'We deal with the local authority; as far as I know we do not deal with any special section of it.'

The Assistance Board role, though limited by local overlaps, was by now more clear-cut in emergency. The man sent down to Plymouth quickly improved local arrangements. 'Being a man of imagination (he) did the simple things which matter, like ending the long and tiring standing queues at the Board offices by simply getting hold of enough ordinary chairs for the homeless to sit on while they waited.' After hours spent with Board personnel, their set-up could be described as 'the most smooth-working, sympathetic and efficient organization of any yet seen in any department of blitz-town life.' What they did was to show concern for the unfortunate, respect for the miserable individual.

During the last week of April this Assistance Board handled, sympathetically, 11,926 Plymouth claimants for compensation for damage or related help. 5000 were given immediate cash advances totalling £70,000. Not a very marvellous standard for the bereft, perhaps, yet by the standard of the moment the treatment looked bountiful. What the Board Chief called 'the dab in the hand', a tenner on the spot, had immense, encouraging effect.

5000 out of 11,926: a fraction again of the affected. Many more left town unassisted. Some were helped by voluntary bodies, though (we found) these had trouble cooperating with each other. This was a very general difficulty. Titmuss considered that even within the Ministry of Health 'different divisions ... were not conversant with each other's policies', covering as they did homelessness, evacuation, rehousing, sewer repairs, hospitals, casualty services and ambulances, corpse disposal, etc. London's Special Commissioner Willink wrote to that Ministry (on 15 April) concluding that failures in coordination between Assistance Board and local authorities 'caused more inconvenience... than any other single factor' in the London blitz; in provincial towns (except Plymouth) likewise.

About the only other item on the human landscape making a markedly favourable impact was Lady Astor, who was here, there and anywhere, talking to everyone about anything, at least once doing cartwheels through a public shelter to shake people up. What did it matter what she said, or did? She was *there*, arguing with Winston, the press, her husband, one of those very rare figures, mostly feminine, who emerge as personalities in reports which tell so heavily of negativity; reports which sought strict anonymity but could not avoid mentioning her regularly by name.

Nancy Langhorne was born in West Virginia, US in 1879, marrying the Lord Mayor-to-be in 1906. She entered parliament for Plymouth (Sutton) in 1919 and would retire in 1945. When the blitz fell, she was 62, a Christian Scientist, feminist, teetotaller, her informal un-British approach distinguishing her from nearly everyone else on a record which usually showed zero impact for local MPs (and Mayoresses) after a blitz. Yet all they had to do was 'show', in the old boxing sense: to enter the ring as a combatant – win or lose, never mind, just show.[6]

Let her husband, Lord Astor, have the last words on his bailiwick. 22 May 1941, eight days after the blitz faded out over Birmingham, he wrote to the Regional Commissioner for the south-west across the moors in Bristol, suggesting from the experience of Plymouth that elderly or ineffective elected councillors, aldermen and paid local officials should be invited to make way for younger, more vigorous men; that local authority boundaries should not at this time be allowed to operate as administrative barriers, alibis for inaction; and that the same service should be pooled regionally in advance of attack. He summed up: 'Local authorities did not profit from each other's experience, neither Regional Headquarters nor Whitehall succeeded in conveying to them the need for a bigger, swifter, more efficient preventive organization... The peace-time system of slow committee rule, of red-tape, of endless letter-writing between London, Regional Headquarters and the periphery has shown itself an absolute danger to human life.' More, though, than life and death were involved; distress, humiliation, fear were products of blitzing, unavoidable but only up to a point. That point was, on this record, passed in Plymouth, as before in other southern ports. Drake's town had been brought very low, despite much courage and the patience of poverty. How far was this so further north and inland, especially in the great industrial cities second only to London?

8 The Industrial North

'We may be in the gutter,
But we have our eyes on the sky.'

OSCAR WILDE
(*Lady Windermere's Fan*)

It is not the object of this exercise in direct observation to spell out each and every blitz situation in Britain. To do so would exhaust both publishing space and the patience of the reader. Rather the intention is to illustrate and document processes of varying 'typicality' under extraordinary, atypical circumstances. In this case those concerned are British. More important perhaps, they are human, *Homo Sapiens*, reacting to new experience with diversity. Although there were extremes in these reactions, in only the narrowest ('Whitehall') sense was it true to say that the 'morale' of Plymouth – for instance – had 'gone'. The loss was much more complicated than that. The men and women fighting in these blitz-battles were fighting to survive and continue. They were not fighting – as at Waterloo or Arnhem – to win. Simply to live on and through the blitz was a kind of victory; much more so to carry on working as well, as many were in effect led by need to do. The victories were made up of minute personal triumphs adding up to massive, adaptative achievement, despite savage defeats upon the way.

Having by now hopefully established some of the patterns for living through the blitz in six places – London, Coventry, Southampton, Portsmouth, Bristol and Plymouth – we can afford to be more selective. Therefore we shall now consider more briefly another comparable series of raids further north before going on to collate the material and seek wider conclusions.

(1) Merseyside and 'martial law'

The twin conurbations of Liverpool and Birkenhead flanking the Merseyside, population then around one and a third million, was Britain's major Atlantic port, start of the essential sea-lane to North America. As such, it received special German attention from the start, sixteen large attacks, eight ranking as full blitzes (only London had more). The most peculiar feature of these bombardments was their ultra-erraticism, usually with a series in close sequence then a long lull, different from anything in the south:

	Month	*No. of main attacks*	*Total bombers effectively involved*	*Number of serial nights (doubled or more)*
1940	August	4	448	4
	September	–	–	–
	October	–	–	–
	November	1	324	–
	December	2	504	2
1941	January	–	–	–
	February	–	–	–
	March	2	381	2
	April	(2)	(135)	–
	May	5	625	4

This kind of pattern made it difficult for people to 'train' themselves effectively. Let us pick up the story after the largest of all the Merseyside raids, 324 bombers on 28 November 1940. A factory worker living in Woolton Road, Liverpool, wrote in her diary:

At 7.30 the siren wailed. It had been three or four days since we had heard it and we had gradually sunk back to our pre-war routine. We were caught absolutely unprepared. We didn't think it would be much of a raid so although I dashed upstairs immediately I still didn't bother changing. We didn't go out to the Anderson until the gun-fire. For the first two or three hours there wasn't much, except heavy gun-fire; clear sounds of planes and the crump of bombs. We all got intensely weary and rather bored, and when there was rather a long lull, we decided to go in to the house. I climbed out of the shelter but the guns started up all over again so I got back in again. We were all very cold by this time but this was our own fault for not being prepared with adequate blankets. At about 12.30, C – our neighbour's son – came into the shelter. (We share the shelter with our neighbours.) He had shelter-hopped up from his girl friend's house, around the neighbourhood of which he had been helping to put incendiaries out by throwing soil on them. The attic of one house had caught fire but they had it out in about 30 mins, by using a stirrup pump. C wouldn't stay in the shelter, in spite of his mother's agitation, but got out and was standing on the front door-step talking to our lodger.

This lodger never sheltered: he liked 'to keep an eye on the house'. Next came two terrific vivid flashes, then two loud explosions. They sat 'stupefied for a moment'. C looked in, asked: 'I say! Did you hear those two!' They were landmines. Presently they all went out on the road to see the flares and fires – 'the sky in one part was a lovely pink shade and the houses and trees were jet black against it.' On the all-clear, 4 a.m., all the neighbours got together, talked, went over and saw the Co-op shop opposite with windows blown in. Back into the house: 'everything covered in plaster and the door of the cabinets and bookcase all blown open. After putting my hair in curlers and having a wash, I climbed into bed at 4.30. I got up at 7.30 and *went to work.*' On the way to work her usual tram was missing, so she was half an hour late, 9 a.m. The last girl did not arrive until 9.30, with apologies to the boss – but she had had to walk the whole way. Several of the girls had been bombed out and came in bandages. Three sisters with whom our diarist worked closely had their brother, his wife and child killed as well as their own home bombed out. The manager sent two of the firm's vans to move what was left of their furniture, while the under-manager took the three to his house, empty as his wife and child were already evacuated.

Three weeks passed before the next significant bombing. Then, just before Christmas, two harsh nights. One by now blitz-wise whole-time observer arrived on 23 December, the rest of the team on 1 January. The place looked at first much as in the south. But within the hour, some fresh human messages were coming through. A 24-page report written 6 January 1941 concludes:

On the spot visibly good morale of Liverpool showed itself in many ways, including:

(i) Relative scarcity of rumour, and no recorded mention of (Lord) Haw-Haw (in contrast to the southern towns).
(ii) No obsessional talk about air-raid damage and many hours of overheards without any air-raid talk at all.
(iii) No visible alarm when sirens sounded, either by day or night; very few people leaving pubs or cinemas, or roaming in the street.
(iv) Only small numbers of people doing sight-seeing tours of air-raid damage or staring at burnt-out buildings in the town centre (– a near obsession, elsewhere, from the start).
(v) Large amount of singing and whistling in the streets – most popular tune the new comedy number, 'Bless 'em all, The long and the short and the tall.'
(vi) Shopping and general life of the town and town centre apparently carrying on much as usual. Reasonable amount of night transport, including some taxis available even in raids at night. Restaurants serving meals up till 9 p.m. and doing good business. Pubs also doing good night business, especially in the 'last hour' normal peacetime pub trajectory, (put forward one hour by earlier closing).
(vii) Several cinemas, the theatre and the pantomime all carrying on after dark, and playing to moderate, though not full houses. But these, and all cinemas, doing big afternoon business, not affected by daytime alerts. Dance halls well filled at night.
(viii) Good humour and laughter, friendly and helpful attitude of police to strangers, etc.
(ix) Low degree of gas-mask carrying (sign of low anxiety).
(x) Conspicuous numbers of children and women. Only a small proportion of those eligible for evacuation have left. The streets are full of children, and they have taken advantage of the blitzed atmosphere. Child hooliganism is quite a typical feature of any Liverpool observation at present. Juvenile crime has not increased but child crime (under 14) is becoming a problem.
(xi) Uniformed armed service personnel conspicuous and conspicuously gay around town, adding a note of revelry – almost a holiday spirit.[1]

In Liverpool between Christmas and New Year you could find 500 in a city centre dance hall with an enemy plane overhead. You did not have to listen to raid talk, day and night. One of us, harking back to University days, likened: 'the centre of Liverpool after dark ... a mixture between a bump-supper night at Cambridge (when Pembroke is head of the river) and a Bank Holiday at Blackpool!' This exceptional cheerfulness was dangerously based on a combination of the post-raid exhilaration more easily developed in the big cities and the usual under-estimate of troubles to come. Yet it brought over a million blitzed people into 1941 in 'good heart'.

This activity was unevenly distributed, however. The young were characteristically more positive in expressing the urge not only to live but to laugh through the blitz. Merseyside's upper and middle classes showed a less positive mood than the rest. Class differences were fairly regularly found in post-blitz situations, insofar as the 'better off' – in the early forties 'class' and income were much more closely linked than in the seventies – of course were better able to buy up rural accommodation, run cars and so on. But the impact of indiscriminate bombing was felt by all. No one thing emerged as exclusive to any one class, group, race, sex or age; differences of personality and of chance always cut across all distinctions based on the more obvious categories. Nevertheless, in Liverpool as again inland in Manchester, the effect of a large industrial upper middle class was felt with a little extra emphasis. In both places, emergency services, manned by volunteers who were mainly derived from this source, were thought to be damagingly influenced by the failure of personnel to report for duty in or after severe attacks.

Dockers (15% of the Merseyside work-force) and Irish (or both at once) were also important activist groups in the forties. Leftism among the former, strong and somewhat 'expatriate' Catholicism among the latter, were minor sores in the body politic. These people did not show any less spirit than others now or at any stage, although dockers remained consistently dissident, and difficult industrially. Maybe the Catholicism offset the Communism in times of crisis?

With exceptions, then, Merseyside responded almost joyously to the pre-Christmas blitzing; as if it had won. The lack of concentrated damage at the city centre helped. But once more one must fall back on local spirit, the unmeasured intangibles, for a full explanation.

Along with a hardy northern tradition of endurance, which had come to a climax within most living memories with mass unemployment in 1931, there went (we noted) 'a considerable cynicism about the possibility of ever getting *anything* improved by political, grumble, demonstration or petition methods'. The look-after-yourself-Jack philosophy was extra-firmly implanted. And the political struggles of pre-war years had left the left with little feeling of achievement. In January 1941 specifically: 'In all classes of Liverpool society, there is a widespread rather despairing contempt for the city council and its alleged corruptions. If anything, this has increased lately, since the city was governed by an Emergency Committee of three, with the shipowner mayor in the chair.'

The public invisibility – so to speak – of this Emergency Committee gave concern. If anything, it 'increased the gulf between the leaders and the led'. We feared that if things got worse 'this lack of confidence might become significant'. Meanwhile, the surest laugh in January was to ask any informed citizen what he thought about the present showing of local government. Jokes about passing the buck were capped by one explaining how the Town Clerk had perfected a technique for passing the buck to himself – in his various capacities as Clerk of the Rating Authority, Education, Old Age Pensions, Public Assistance, the Licensing Authority, Commissioner of Income Tax, Clerk to the Mersey Tunnel, Food Controller, Light and Fuel Overseer and of course co-ordinator of all ARP services. He was inevitably called 'Pooh-Bah' too.

Personalities – or to be more precise impersonalities – apart, the main material cause for concern was the local shelter system. There were too few. Surface brick shelters were doorless, bitterly cold in the northerly winds. Both in Liverpool and Manchester large basement shelters in stores and large buildings, holding from 200 to 6000 persons, had been developed as a concession to the demand for deep shelters. An independent survey during November gives the picture:

> Comparative study of shelters in Manchester and Merseyside showed that in basement shelters eight out of twenty-three were heated in Manchester, two out of thirteen on Merseyside. Ten out of twenty-three were reasonably well ventilated in Manchester, four out of thirteen in Liverpool. Four out of twenty-three Manchester shelters could provide warm drinks, but only one out of thirteen in Liverpool. One of the Manchester and five of the Merseyside basement shelters were very damp.

Thus, all round, the Liverpool basement shelters, nearly all large ones, compared unfavourably with those in Manchester. A striking feature of all the shelters was the total absence of a first-aid chest. In one case, a shelter accommodating thousands of people had no first-aid post and no nurse. A skilled social worker gave his own careful impression thus: 'Improvised arrangements, judging by conditions revealed on Merseyside, have created living conditions for many citizens which abolish most of the improvements in sanitation, cleanliness and health, made during the last century. The squalor revealed in some of the shelters visited was almost Hogarthian.'

Large shelters thus surveyed catered for 26,975 theoretical users. Most had three-quarters of their height above ground. Many were really halls, with no blast walls; two were under town halls in the centre. In the absence of anything better, people from the crowded 'slum' districts were in regular occupation of the safest-looking – 'after a long succession of undisturbed nights' some were full nightly. One better fitted-up Mersey shelter, a private development, let space for up to 27s. 6d. a week. The Lancashire and Cheshire Community Council, privately circulating available data on shelters and Rest Centres, expressed, discreetly, grave alarm. They noted the extent to which lessons learned in the first months of the London blitz were apparently not being applied in the great northern cities.

With no big attacks in January and February 1941, spring came to Merseyside (even more than to Plymouth) as a hideous shock with mid-March's resumed blitzing, running on with five consecutive nights in one week of early May. Look now for a moment at the great port straddling the wide river, through the eyes of an observer returning there, in May. A member of Mass-Observation's regular blitz team, this young man had spent part of his childhood on Merseyside. His family were intimately connected with the city and he had friends in the university, Conservative Party and social welfare systems, with an uncle high up in ARP.

This man found those January predictions had moved into reality. Public shelters were now discredited. The Rest Centres had proved quite inadequate, in some instances 'hopeless'. Emergency feeding arrangements had virtually collapsed. Mobile canteens were once more conspicuous by their immobile inadequacy, despite rumours that the authorities had refused outside offers because all was 'under control'. No hot food was available to most inner-city dwellers for days. Entertainments folded. Some pubs which stayed open ran out of beer. There were no cigarettes. Public transport was in chaos. 'An atmosphere of ineptitude seemed to oppress the town.'

Groping for words to generalize the obscure, the May report concluded: 'The general feeling – it is difficult exactly to express it, but residents spoken to felt it too – that there was no power or drive left in Liverpool.' A sense of damp defeat lay across the scarred city scenery. Little was being visibly done, by mid-May, 'to put people back on their feet after the worst continuous battering any people yet had'. What could only be described as 'unprintably violent comments on local leadership', were to be heard, including some from sources which had been at the worst cynical on previous visits. There appeared by this time to be an 'almost complete divorce' between a few key politicians and officers at the top and the worried, bewildered 99%, with no serious attempt to keep the 99% informed about either present action or future intention. However impossible the actual situation, this total failure to communicate, this silence, seemed inexplicable. Nevertheless, rumour had it that the city's surviving marble halls echoed with talk of 'honours', OBE's.

This failure in leadership closely resembled Southampton's, but on a much larger scale (nearly ten times the population). Liverpool's dissatisfaction in May reached a level of 'vehemence' (our word at the time) unmatched elsewhere. (In Southampton and in other cities criticism of the authorities had been slight and slow to grow.)

Here, too, for the first time in any blitzed town, there were publicly expressed signs of disgruntlement moving towards a willingness to surrender. This was fanned and inflated by a great growth of rumour, that safety valve of anxiety. There thus emerged one of the great rumours of the war, flashed right across the country after Merseyside's last big night on 7 May. The 'Martial Law' rumour reached the proportions of the 'hairy nun' the year before. The nun in question, 1940 heroine that never was, entered a railway carriage at some intermediate station between

Brighton and Victoria – or between Plymouth and Exeter, Wigan and Manchester, Bootle and Warrington – to sit down comfortably in the nearly full compartment. After a while, though, her calf started to itch her. Thoughtlessly, the devout lady lifted her robe to have a quick scratch. Horror to behold: 'her' leg was darkly matted in manly hair. It was a parachutist.

The Liverpool rumour of martial law was encouraged by the police and army temporarily closing the city centre to cars and persons without special business, in order to ease congestion and enable surface debris and unexploded bombs to be cleared. There was no public explanation for this necessary measure. Some of the victims jumped to conclusions. The conduct of others added fuel. The story spread fast, was picked up within hours in London; from a responsible MP, a BBC official, a senior civil servant, an important editor: all affirming a peace demonstration had been staged – the long awaited revolution in embryo. So: Liverpool had been put under martial law.

Impetus was added to this distortion by the difficulty of getting letters out of town, the refusal of all telephone calls and telegrams, the glaring paucity of factual information. Yet as late as Sunday, 18 May, with the rumour rife outside, pedestrians were being allowed all over, and we found the central streets absolutely congested with sightseers enjoying the mess.

A Woman Auxiliary Air Force (WAAF) girl stationed between Liverpool and Manchester reported on what was going on at nearby Preston, a large cotton town. She wrote (10 May):

Preston, so far, has had no bombs. There is a certain smugness here, among one section of the people, but others hearing the most heart-rending tales of Liverpool's suffering, are inclined to think: 'Our turn may be next.' In the whirl of WAAF life, I haven't had a chance to think much about bombing beyond a sympathetic 'London got it badly last night' or 'Liverpool's caught it again'; raids have scarcely been on my mind until the day after the worst Liverpool blitz. Then, with natives of Liverpool in the WAAF tearing hair out for their relatives, I began to realize the seriousness of the position.

All sorts of rumour circulated about Liverpool and martial law!

Jean: 'Have you heard about Liverpool?'
Me: 'What about Liverpool?'

Jean: 'Everyone's talking about it. Surely you must have – they say (she whispered this bit, dramatically) they say the people there want to give in.'
Me: 'I don't blame them, judging from the reports of the raids there.'
Jean: 'Yes, but I don't believe it's the people. I think it's those wretched Irish trying to create panic. It's very easy to do. They're going around shouting "Stop the war" and "We've had enough"! *English* people wouldn't do that.'
Me: 'It's surprising what you'll do when you've lost everything.'
Jean: 'I was told they've got martial law there, and that if anyone is found saying they want the war stopped, they're *shot on the spot*.'

Later that morning, our observer and friend Jean hitch-hiked on a lorry that had come from Liverpool. Jean was anxious to convince herself that things were not as bad as she had heard.

Jean: 'Some of the stories are too ghastly to believe.'
Lorry Driver: 'However bad they are, they can't be worse than the truth. That's a fact. I've been doing what all the other lorries have – taking people out into the fields, and just leaving them. The local council hasn't done a thing – only given them food. That I will say.'
Me: 'They must have known it was going to happen.'
Driver: 'You bet they did. They'd got the coffins all made out last spring – 50,000 of them – that's why we were so short of wood. But not a jot for the living. There's 50,917 dead, and God-knows-how-many wounded, just walking the streets, with their bandages on.'
Jean: 'There's martial law, isn't there?'
Driver: 'Well, not exactly. But there's a lot of military with bayonets – they've more or less taken it over, as you might say.'
Me: 'How are the people bearing up? Is the morale good?'
Driver: 'They can't help themselves.'
Jean: 'Well, you've relieved my mind ever so much. I was imagining all sorts of things.'
Driver: 'You can't imagine anything as bad as it really is.'

Much further away, over in unbombed Staffordshire, a Leek working man sent in his observations dated 17 May: '*General Morale:* Very unsteady. This has been a week of gruesome rumours which were briefly as follows – (1) Train loads of unidentified corpses have been sent from Merseyside for mass cremation: (2) Martial Law has had to be put in operation in several heavily raided industrial areas: (3) Homeless and hungry people have marched around in bombed areas, carrying white flags and howling protests: (4) Food riots are taking place.' So this rumour magnified 'reality' and became 'fact' over much of Britain. It has survived as a belief into post-war.[2]

Merseyside took a long time to recover confidence in its leaders. A year after the blitz ended, in May 1942, a report based on a ten-day visit began by noticing that what struck the outsider first was an air of lassitude, although for months there had been 'no big impact of war'.

A month before, in April, another aspect of this mood was expressed in a parliamentary by-election at Wallasey where a local journalist, George Reakes, expelled from the local Labour Party for supporting Chamberlain over Munich, won the 'safe' seat with a huge majority over the official candidate, put up with both National Government and Communist Party support.[3]

An observer had a 1942 May Day conversation with a fire-watcher across the river at Birkenhead docks, and considered it 'not untypical':

Observer: 'What do you think of the war?'
Fire-watcher: 'It's ghastly.'
Observer: 'What's wrong with it?'
Fire-watcher: 'It's bloody awful.'
Observer: 'What should we do about it then?'
Fire-watcher: 'Have a change . . . Start again. Them running it has got the wrong end of the stick. We want some young blood, some fresh young blood.'

(2) Manchester and Birmingham inland

'Manchester was such an unco-ordinated, overlapping, jumbled-up place, that even at the best of times "Manchester feeling" and a positive Manchester outlook were liable to be lacking.' Thus a Press Association (Special Report Unit) paraphrasing an old Mass-Observation blitz-report, for a 1973 Granada TV treatment of what happened in the Mancunian raids of winter 1940–41! Here 730,000 people – overlapping into Salford and the wide industrial belt towards the cotton towns like Bolton, all about – received only three fully blitz-scale attacks, two immediately after Merseyside on 22–3 December (doubled with 441 planes) and one for the New Year on 9 January (143). These three put Manchester eleventh on the blitz ratings, but in terms of human reaction they may have rated higher. Mancunians got very upset, at a time when Liverpudlians were still pretty confident. So it was that early in January – before the ninth raid – we remarked: 'Going from Liverpool to Manchester was like going from an atmosphere of reasonable cheerfulness into an atmosphere of barely restrained depression. Directly investigators got into the town, only an hour away from Liverpool by road, they felt themselves back in the blitzed town atmosphere with which they had grown familiar in the South.'[1] Large, straggling, inland Manchester, differing from the clear-cut coastal cities, offered no specially identifiable targets beyond the human mass. In this connection, it is well to recall that on 8 May 1941, after eight months of night bombing experience, Nottingham was raided by 95 German bombers actually targeted for Derby, fifteen miles away. The Luftwaffe crews reported success. In consequence, other crews briefed to bomb Nottingham itself dropped their loads on open countryside as far east as that place lies east of Derby.

The Manchester double blitz fell on the Sunday and Monday nights before Christmas. (The Germans gave Britain a brief break for 24–6 December inclusive.) The timing 'gave a tragic bitterness to the whole affair'. The bitterness had ironic side-effects. One: many better-off voluntary helpers in the social services had taken the week off from the Rest Centres and other emergency systems.

Another: there were no newspapers at a time when the local press could have done something to rally people – who, as usual, were enraged at the radio underplaying their suffering. When the (then *Manchester*) *Guardian* reappeared, it glowingly but inaccurately described the Christmas turkeys supposedly served at each Rest Centre (in fact at only one).

As it happened few were talking turkey that Christmas. Instead, along with inevitably obsessive bomb-damage talk, rumour ran free, including a rich set attributable to Lord Haw Haw on the German radio. One of the best native efforts was an updated variant of our old friend 'the hairy nun', whereby a parachutist had landed dressed as a nurse and was now working in one of the city hospitals. In the streets there was 'much silence and a great night silence'. Night-life closed down.

The presence of the Regional Commissioner's office in Manchester did not in this case produce much effect, partly because of the immediately preceding major raids on Merseyside in the same region; partly, also, owing to a curiously detached attitude shown by that office in public. On 23 December the Regional Commissioner issued a roneo'ed folio sheet listing the points he evidently considered crucially important at that moment. This started with three demi-commands phrased officially:

1. Do not congregate at places where the civil defences are working.
2. Be back at home before dark.
3. It has come to notice that certain buildings the occupiers of which are required by law to provide fire-watchers have failed to do so. Further neglect will cause the law to be enforced most rigorously.

People patently ignored 1, didn't need 2, while 3 was belated indeed – and put harshly in a jargon unsuitable to the occasion.

The same morning the Emergency Committee of the City Council (see below) issued a lavish letter addressed to the CITIZENS OF MANCHESTER. These gentlemen devoted one large paragraph to praising every corner of officialdom for 'magnificent' service the previous night, in a style familiar to observers ever since the start in Southampton. A much shorter, one sentence paragraph then expressed admiration for the 'fortitude and cheerfulness' with which the common citizenry 'met this ordeal'.

The centres of municipal power radiated this sort of self-satisfaction. But they had in fact been moving towards reorganization in a major effort to face up to the problems, as indicated in the Mass-Observation report after Christmas:

One of the most important factors of all in Manchester, is the bad organization for dealing with the results of a raid, especially with the numerous homeless. Post-blitz morale considerably depends on the way in which the human problems are dealt with. Those who are not bombed out are affected by the morale of those who are. The morale of the bombed out largely depends on the care they get in the first thirty-six hours, of how many are steered into Rest Centres, and of how sympathetically they are treated there. The Rest Centres in Manchester were almost as unsatisfactory and unprepared, in some cases, as those in the East End three and a half months before. They would have been much worse had it not been for a determined attack by the Local Council on the constitution of the Emergency Committee of three, which led to the Chairman resigning, to the Committee being extended to six, and consequently to an Air Raid Welfare Officer and a whole new department to take over responsibility for Rest Centres, etc. This occurred only a few weeks before the blitz, and the new department, while it had done much, had obviously not been able to do everything.

Terrific efforts (we wrote) had been made in the earlier part of December, stimulated in part by the joint shelter and centre survey already mentioned for Liverpool. The serious inadequacies revealed by this independent study could not be quickly corrected where they involved such troubles as badly prepared basement shelters. On the human side, change was made easier by an unusually strong group of liberal dons, students and social workers (associated with the university in an old local tradition).

Nevertheless, a typical Rest Centre still had only one blanket per stretcher bed. Camp beds in three others were 'due to be returned to the manufacturers because they were falling to pieces'. No centre had an adequate first-aid chest, though there was plenty of spare clothing from U S gifts.

A Bolton coal-man working around Manchester visited three of their Rest Centres. He judged 'the misery and despair' of the people therein as 'past description'. Not quite as bad as that, perhaps. But what, one amazedly asked, of the lessons of London, of Coventry and all the rest? Why had this great rich city not learned before?

A clerical worker, 20, her husband soldiering in the Far East, ended her diary that Christmas Eve by reproaching the BBC for its underplaying of the local blitz, then went on: 'You will hear a lot of talk about Manchester carrying on. I suppose we are ... but as one who lives here, it's a rather weary carrying on. We are carrying on' (she concluded) '*because we've got to.*' And as an afterthought she added, in some surprise, that she had 'not heard a single person speak in favour of reprisals'.

Manchester faded out of the blitz scene early in January, shaken but not near the stage of generating its own 'martial law' excitements. Birmingham, larger – second largest city in England with over a million inhabitants – more accessible, more important in the armaments industry, had the advantage of size as well as a longer process of blitz familiarization. It had also an elaborate civic pride; in those pre-Granada days certainly stronger than Manchester's.

Beginning after adjacent Coventry; 19–22 November with a triple-night blitz (totalling some 677 bombers, with over 450 dead); again 1–3 December (but only 103 planes); 11 December (a raid lasting 13 hours); 11 March, 9–10 April (443 bombers); and at the very end 16 May (111): Brummagen people had a wide spread of the pain. Altogether there were 73 separate raids, large and small, delivering 5219 recorded bombs, of which 930 did not explode. That meant about one bomb for every ten acres of the city, the same as for one month on Chelsea or Islington, near the top of the list for London's boroughs.

A striking material feature in Birmingham was the extent of fire damage. The inadequacy of precautions here raised an unusual (for wartime) civic storm as late as 6 May 1941, when a Labour councillor demanded a private meeting of the City Council. This was not held until 29 May, when the meeting passed a (secret) amendment calling for some sort of regional set-up to supervise the fire services.

The local official historian, T. A. Sutcliffe, has concluded that generally speaking Birmingham's 'came to be recognized as one of the most efficient civil defence organizations in the country'.

But his supporting reference is an 'unpublished account', without source or date.[2] A different emphasis was given by the Research and Experiments Department in the Ministry of Home Security, working soon after the blitz to assess the effects in terms of implications for possible future assaults. RED, in 1942, considered that more heavy raiding could have led to 'a dissolution of the civic entity' and rendered Birmingham virtually uninhabitable. Vulnerability to incendiary bombing, due to inadequate staff and bad co-ordination, was especially stressed. 90% of the industrial damage had been caused by fire. 'Relations between certain sections of the civil defence services, notably the fire brigade and the Auxiliary Fire Service, left much to be desired.' On the other hand, the effect on industrial work was 'relatively small'. Whenever possible, people went on working.[3]

Birmingham, the RED men concluded, was not only caught unawares that winter, the December raids disclosed 'grave deficiencies' in the local system. The authorities failed to help the victims of bombing promptly or adequately. In one place, the bombed took over the premises of a transport depot; there were 'disorderly scenes'. Rescue workers, overstrained, were 'going crackers'. Looting became serious.

This RED study had the disadvantage of being carried out mostly by interview, a year or so *after* the events. Memories were by then distorted. But the tone of criticism is strong, based as it was on continuing fear by some responsible leaders that the blitz might be resumed. The work of top-level scientists inside government, the report goes considerably further than Mass-Observation's observations into January 1941. RED considered rational local demands for better relief services, a better warning system and so on.

Another reaction comes from Dr Sutcliffe's history, already cited; he writes: 'The Luftwaffe, although sometimes portrayed as a bestial monster, more often appeared in Birmingham as a very tame and ragged eagle, which lost its way, dropped its bombs at the wrong target or no target at all.' As a result, this historian argues, most Midlanders realized that the blitz was random rather than deliberately vindictive: 'even in the heavily bombed area of back-to-back housing in the inner ring little violent hatred of the enemy seems to have been aroused', he says. There was 'an absence of any serious bitterness against the raiders'. This was also Mass-Observation's experience, not only in Birmingham.

Misadvised by loud press cries, especially from Churchill's War Cabinet colleague and friend Beaverbrook, a doctor stood as a 'Reprisals Candidate' in a parliamentary by-election at King's Norton, Birmingham, in March 1941. He forfeited his deposit, as did his Pacifist opponent. The Conservative, an ex-army captain, got 21,753 votes to their combined 3248. It became the done thing, through published opinion, to shout for vengeance. To call for peace was treated exactly the other way round, as treacherous. Neither campaign ever enlisted more than a fractional minority of Britons even in the worst of the blitz.[4]

For through thick and thin, most people kept as much as they could of their ordinary cool in extraordinary times. In mighty Birmingham, a high proportion managed to stay put, whatever the difficulties. It is well to bear this constantly in mind. Go back for a moment to the diary of a housewife, Mrs S, reacting to 704 bombers on the third night of the city's first big assaults (19, 20 and 22 November):

We rushed down to the shelter, the gunfire was hot and the effect of the fires in the triangle of windows facing us was terrifying because of the thought of all those houses on fire and we in the centre.

We arranged ourselves as we could, Les bailed, the bombs were terrific, I covered G up as much I could to save her ears, she could hardly breathe for bronchitis.

The tension, the waves, good Lord I thought they would never stop, not 500 planes came over but a 1000 I should think. G eventually dropped asleep from sheer exhaustion and stayed asleep till the All Clear, I was grateful for that.

Mr Brown from next door came down during the gunfire, he was afraid and he didn't mind admitting it, it didn't stop him from helping to put the fires out though nor bringing us a cup of tea down. It was quite a relief having a chat to someone but he thought we had dozed off later, so he didn't come down after midnight. I dozed off, mind you.

In this Anderson Mrs S thought she heard screams among the bangs. Next morning she learned they came from a near-by public surface shelter after a scare over 'time bombs'. The warden had made 700 people evacuate in the dark half-clothed and 'nearly mad with fear' and move to the Co-op repair shop for safety. She herself edged nearer to the extreme experience when she learned also that a neighbour and her cousin were killed by a direct hit. A woman friend told her how they identified the bodies: 'It was a ghastly experience. Her husband had to go to various undertakers to wade through body after body and, not having seen their cousin for some time it was impossible for them to know what she was wearing.' Mrs S talked to a woman air-raid warden friend about the goings-on with the local undertaker, and reported her as follows: 'She thinks *the* most horrible job of all during war time is the undertaker's. Her brother-in-law is an undertaker, he has had 7 beautiful horses bombed and when one of the B'ham raids were on he had 46 sacks delivered to him with the dead remains of victims. He had to clean and straighten them up as well as he could and lay them out for identification.'

(3) The Clyde towns (with two women especially)

In Scotland, only Glasgow (population *c.* 1,100,000) with the associated Clydeside shipyard and industrial towns, received full blitzing, as the north-western extension of the Luftwaffe's main effort. It was left almost unmolested all autumn and winter, to receive all its five big raids between mid-March and mid-May, with some 1300 tons of H E. The Clyde towns thus had an extra large margin of time to prepare, after the blitz had begun further south.

Mass-Observation was fortunate here, in that we were sent to the Clyde to make an intensive study of the industrial situation, submitting a report dated 7 March, six days before the major bombardment began. We were right there, therefore, when the skies opened on a largely unexpectant Clydeside, still full of that industrial unrest for which it has been even more famous than Merseyside. Immunity from bombing had not improved that atmosphere. Some leading citizens argued that 'what Glasgow needs is a few bombs'. We wrote of this argument then:

If this is accepted, then leadership and propaganda seem pretty meaningless words. And it is *extremely unlikely* that the Germans will, after leaving Clydeside alone for months, kindly oblige by dropping just enough bombs to stiffen morale. Indeed, a number of important Glasgow people held the view that the Germans are leaving Glasgow alone and will continue to leave it alone precisely to produce the effect which is now being produced, namely of gathering unrest and selfish interest. It is not that Clydeside workers are against the war or for peace. They want to win it as much as anyone, though there is a considerable Maxtonish minority. It is rather that Clydeside workers are *also* having a war of their own, that they cannot forget the numerous battles of the past thirty years, and cannot overcome the bitter memory of industrial insecurity in the past ten years and their distrust of the motives of managers and employers. Bombs now might well focus the hatred outwards. Much would depend on where they fell and how the subsequent problems were dealt with.[1]

A large chunk of the Clyde's population of nearly two million had, by March 1941, wilfully put away the possibility of being blitzed. A snap survey gave:

30% expecting heavy raids (about half of these 'soon')
28% not expecting anything of the kind
42% vague or indifferent, largely ignoring the whole issue.

Some said it would be better to have a big raid and 'get it over with', like this elderly housewife: 'I wish to goodness it would come and *be done with it*.' Many and varied were the explanations for non-attack, sometimes vehement. Nine favourite themes illustrate the diversity within even the smallest of the three groups shown as statistics above:

(i) air pockets over the Clyde generally;
(ii) mountainous area too dangerous for night flying;
(iii) a magnetic element in the mountains, which dislocates aircraft engines;
(iv) impossibility of locating the Clyde in a network of lochs and sea;
(v) adequacy of AA defences and depth of guns on the periphery;
(vi) distance inland or overland (very popular);
(vii) too far from German bases;
(viii) Germans not antagonistic to Scotland;
(ix) Germans believe revolution will develop here so long as bombs *don't* stir up the people (common upper and middle-class opinion).

These canny Scots had built themselves something of a fool's paradise, free of anxiety as well as preparations. Only one in ten said they had made any actual plans for a blitz-scale emergency. Fortunately for the other nine the city fathers had used the long lull to advantage. An energetic Lord Provost, the elected head, was himself a sunshine spokesman, full of optimism in public, but his staff were notably alert. The Regional Commission was also strongly represented in Glasgow by an able ex-industrialist.[2]

A by now very experienced viewer of blitzes commented that 'the whole experience of moving in official circles in Glasgow was something quite new'. Liaison between departments appeared intimate, 'much closer than found anywhere else'. The emergency feeding plan, so inadequately prepared down south, looked good here, with ten peripheral cooking centres and with mobile canteens ready to be mobile. The Education Committee had prepared cooking facilities for 20,000. Rest Centres numbered an impressive 102, not yet all fully equipped. Standard issue included babies' bottles with teats, 250 cups compared with only 6 beds, 18 hurricane lamps, 1 box of matches (no cigarettes), 24 napkins (but no toilet paper), a tin opener, 2 enamelled pails, a dozen safety pins, 5 lbs of sugar and 1½ of tea with one 8-gallon and one 4-gallon tea urn. Nothing wonderful – but ample compared with most of the south.

An unusually elaborate plan to move people on from one of the Rest Centres into prepared, indexed billets *inside the area*, covering a potential 54,000 people, was aimed to offset trekking – and helped to do so. Assistance seemed in good shape, too, starting out with one of this department's fresh looks at its role (in the style seen at Plymouth). Only air-raid shelters were regarded, early in March 1941, as still very unsatisfactory. The main trouble here was physical, location and structure, not readily to be redressed by better organization, co-ordination or information.

In the event, these preparations did enable the area as a whole to support the mid-March to May attacks with less distress and dislocation than might have been expected, especially in view of the political and economic divisions. The lack of public expectancy did not prove so disastrous as one had feared, because it had not infected the responsible authorities to the usual extent. This great scattered conurbation had enough sense, skill and room for manoeuvre to adapt and improvise so as to offset the worst impacts of the unexpected, by now so familiar from smaller towns. But, alas for the best laid plans of modern men, the first big impact of 439 planes on two successive nights fell most heavily on the periphery, notably at the smaller, connected shipbuilding community (47,000) of Clydebank.

Clydebank, which bore the brunt of that first mid-March attack, had been largely evacuated of women and children in 1939. Most had since returned to seeming safety so that over 90% of the children were back in the 12,000 tenement-type buildings. Only seven of these emerged wholly undamaged on 15 March; one third had been demolished, another third rendered at least temporarily uninhabitable. 528 Clydebankers had been killed, 617 badly enough injured to go into hospital (on just under 500 tons of HE). The town's nocturnal population dropped from 50,000 to 2000 overnight. Noting this, Titmuss drily remarks: '*Where they all went to no one knew*.' Once more, the cat escaped the bag. We found thousands of them at first camped up on the moors in a spring reply of Plymouth's Dartmoor exodus.

One of us, fresh from researching Clydebank's industry, got back there while the first raid was still on and continued throughout the second night. He noted some instances – during the actual raids – of hysteria and panic among the crowded population. Unpreparedness had led to 'gross neglect of fire-watching in many streets', markedly contributing to the devastation. Fires caught and spread unopposed.

But in the morning the official machinery proved impressive. Particularly help with emergency evacuation went into immediate effect. In this and related activities the Ministry of Information loud-speaker vans – so futile in the south and midlands – were used slowly, clearly and with effect. And there was little industrial dislocation: surprising in view of all the damage. As in Portsmouth, trekking left a high proportion of the workers with enough sleep to continue working by day. Many Clydebankers worked in the great John Brown shipyard. Of these 'a good proportion' turned up as usual. The 'vast majority' of absentees went back after an average 11–14 days.[3]

The rest of the Clydeside raids, 7 April and 5–6 May, raise few new issues. But the latter assault included heavy bombing of outlying Greenock. Glasgow's jolly Lord Provost had earlier suggested Greenock would never suffer because of its foul weather, the joke going: 'It always rains except when it's snowing.' Some locals afterwards claimed that, if the Germans had concentrated more on Greenock, 'people were getting so they couldn't stand it'. The Germans however left the Clyde alone after 6 May, as all Britain after the following week. And Glasgow proper was still in quite good shape.

There is always the danger that one loses the individual for the general in an analysis of this kind. Before leaving the blitz in the far north, two diarists, differently involved in the Clydeside sequence, will reset the personal tone not only there but for the country as a whole. At this degree of intimacy the word 'typical' is no longer suitable. No one is privately typical of anyone else. These are just two individuals, both women near middle age, the first a spinster, the second a mother of five.

Miss O wrote two pages on the events of March 1941. She lived with her old mother (74) in a Glasgow tenement where the nominated shelter was in the strutted 'close' hallway or basement. Miss Ellen Wilkinson, as number two to the Ministry of Home Security, had inspected shelters in February and criticized the strutted closes, though congratulating the City Engineer for performing 'not only a good but important job'. Being a minister she could not help but echo the persistent Whitehall obsession with shelters. As she told the *Glasgow Herald* (20 February): 'We don't want our people to have that Maginot Line complex.'

Miss O had been writing the previous day's diary at 9 p.m. on 13 March. As she closed the page, she commented: 'What an entry, tinned soups on the day when Merseyside has had the worst blitz for some time. To such an extent do we live in the present.' The previous night she had hardly slept, because a full moon made her expect a raid. Now she was tired and a little sick. For that reason she did not go out as planned. At 9.20 p.m. the sirens sounded. 'The blitz did arrive at Glasgow, and made up amply for its delayed appearance,' she wrote next day; first:

I ran down to Mrs S and Miss and Mr C (about 25 and 17 respectively) our new neighbours, came out too. We got the ladder adjusted to the loft, and carried the stirrup pump, sand and water to the top landing. We opened Mr M's door with his own key, and we opened Mr F's with our door key, which happens to be a master key to practically all the doors in the tenement. Mr F's fire was burning so he must have been expecting to return, but he never did make an appearance.

There was much speculation as to what to do. The rules are that the fire fighting party stands on the top landing, but the C's thought this much too dangerous, and I knew I could not stand there a great length of time without becoming exhausted. Mrs S found she could not either, for Mr S had 'insisted' on her staying below. Later in the evening she came up to tell us her difficulty. When he is nervous the paralysis in his arms gets so bad that he is helpless, and then he gets into a panic. We got back to our own houses and there was a certain amount of going and coming during the early part of the evening.

Our diarist lay in her bed in the dark: 'heavy coat, stole, beret, shoes, gloves, all on ready to go into action.' Mother settled down comfortably in front of the fire, knitting. Mrs S came round at 11 p.m. when there was a great deal of noise outside. 'There will be no bombing' flatly declared Ma: the guns were only practising! But Mrs S thought some of the big flashes must be distant bombs.

At 11.30 there were two intensely vivid flashes, and I thought, 'Crikey me, this one's ours.' And then there was a din and the house swayed to and fro and fro and to, and this way and that. And what a crashing of glasses somewhere outside. I jumped off the bed and ran in mother's direction, and we met in the hall for she was running towards me. She had at last come round to the knowledge that bombs were falling, and looked a bit taken aback. I put my arms around her and said, 'That one was not for us,' and then the second one went off, just a shade further off.

That second one finally disturbed Mother: she changed her carpet slippers for shoes, got her hat and coat, so as to be able 'to make a dash for it'. Then she thought of the 'wee birdie' in the dining-room window – and fetched their pet into the hall, along with the deed box, the cash box 'and of course our handbags'. 'So there we were, all dressed up and nowhere to go.'

Things began to settle down again inside; uproar outside, however:

So I said to mother, we had better sleep in the hall and got two comfortable chairs and the quilts off the beds, and the hot water bottles out of them, and we settled down in the hall.

At 1.30 I thought things better and went to bed, fully clothed, but under the bedding. At 2.30 mother did likewise.

From 1.30 a.m. she heard nothing more except the clock, though no one will believe her on this next morning, as the uproar continued: 'The last time I heard the clock strike was 4.30 a.m. and at 6.30 the all clear woke me from a heavy sleep, and I did not half feel annoyed. I thought I should look up our neighbours now and walked through the building, but never saw a sign of movement anywhere, so I just returned to the house, took off my shoes, corsets and hat, and tried to get back to that land of dreams that I would fain have remained in.' The family emerged to broken glass, the inevitable engagement with post-raid damage which fills page after diary page. The to-and-fro bangs turned out to have been landmines, one of which shattered a garage full of cars, the other flattening a four-storey tenement. Miss O rallied to go to work as usual; finding the office full of rumours phoned in from outside. There were two alerts during the day. Back home, Mother had been making her 'usual shopping round', talking to everyone. She thought their doctor was drugging an elderly nephew who slept through everything last night, except once to wake the family and ask in the middle:

'Why are there so many people moving about?'

'There's an air raid.'

'Is it bad?'

'Very bad.'

'Then light the kitchen fire and invite them in.'

Everyone of course expected the Jerries back again that night. They were not disappointed. One lot of neighbours had already evacuated themselves. But Mr F, missing from his fireside the previous evening was back, 'very nervous'. The diarist livened up her bed-clothed routine:

I was very tired and got the bedroom ready at 8 (i.e. hot water bottles, alarm clock wound), and said I was going to bed 'till it started'. Mother said 'keep your clothes on', and I said, 'no, I prefer going to bed in my night attire'. After all, you don't know that the Jerries are coming, they may not. However, I will put on my best nightdress and bed jacket, so if I am found dead, I'll look respectable. And there I was admiring myself in my finery when the siren went (at 8.30) and I just had to put on the clothes I had just taken off without so much as a split second in bed.

Big noises soon resumed. But Mother sat on knitting beside three windows and the fire. At 11.30 the place shook and she was made to sit in the hall, away from the glass. In return she insisted on light to knit by, whereas her daughter wanted darkness for sleep – so back she went to her bedroom. Miss O recalled some of her dozing thoughts, the last one quite profound in that context:

Here are some of my thoughts. Conflict in the outer world does not disturb you so much as conflict in the inner world of mind. In fact, it is almost beneficial in that it distracts your thoughts from your inner conflicts. What a lark that I with my long record of insomnia should get so much *more* sleep in a blitz than people who have been sleeping well all their lives. Believing wholeheartedly that this nation must stand firm and being willing to pay the price goes a long way to cast off from you the fear of suffering. The pity is that casting fear of suffering out of yourself makes you unsympathetic to the sufferings of others.

Miss O wondered if she should allow herself to sleep so much, in case of fire-bombs. She decided she needed the sleep in order to be efficient by day. The whole experience over two nights was likened to having a tooth out: 'your jaw is frozen but you feel it too.' And when it was over, just relief. Along with this, more exultantly:

The feeling that it was perfectly wonderful, to be on the right side in this stupendous conflict between good and evil. *Oh! I am so glad my lot fell to the twentieth century.*

However, at midnight I went dead asleep. At 4 I woke to find everything quiet, and went to sleep again. Mother says that I slept through some terrific gun firing and how I managed to do so is beyond her.

She awoke to find Ma had gone up to bed at 1 a.m., before the all-clear. Two nearby houses had been demolished overnight. Back at work once more, the staff were all on time. Her special friend Helen arrived in a 'very overwrought state'. Miss B 'seems utterly exhausted from two nights sitting up in a chair', but recovered next day when she reported she fell asleep beside the kitchen sink and was awakened by a 'deafening, unnerving sound ... and what do you think it was? – a drop of water from the tap'.

At lunchtime Miss O found the bus *stacked* with bags and cage birds. This was her first sight of trekkers and evacuees. 'Scores and scores of people were clearing out.' She worried lest the departures might undermine fire-watching. This night was quiet. The diary made up by going on and on about details of bomb damage, incidents, rumours, mistakes, curiosities of the blitz. No criticism of the authorities; no reprisals wished. Over all, hung the persistent patina of bombing past, wryly noted as: 'Post blitz news! The air must be full of dust and soot. Every piece of furniture is coated with dust so thick that you could write your name on it, and when you have dusted it, you go back in half an hour and find it as bad.' It was only after the March blitzing had passed into dust that Miss O realized that most of the worst damage had been further down, on Clydebank – now, she found, an empty town. At the same time, she was learning to feel confidence in her own and her old mother's ability to withstand those Jerries overhead. Others, she thought, seemed more afraid than she – many found strength in themselves by seeing weakness in others. She came to see some of her companions as (comparative) cowards. Yet those lesser braves stuck it out too, sheltering downstairs without any special protection or officially supported shelters. Miss O was particularly aware that no one in the block was any longer prepared to fire-watch upstairs. 'In simple terms,' she concluded, 'they are as yellow as yellow could be.'

The near-consensus was favouring the view that there would be no more big raids anyhow (three were to come, 7 April, 5–6 May). Mrs S, supposedly warden for the block, went away on this pretext ('there is not the *slightest* chance of Glasgow being blitzed again'). Mother O was similarly convinced: 'not another bomb will fall on Glasgow.' She was acid to the grocer that afternoon.

Grocer: 'Next week is full moon. We shall have it again. This time they will concentrate on Parkhead.'
Mother: 'Has Hitler written to tell you?'

On 5 April Miss O, who liked the film *South Riding* at the Cosmo, summed up an 'uneventful week'. Like her grocer, she felt the coming full moon didn't look so good. Early Monday morning 7 April, her diary started: 'So far a most uneventful day.' Then she devoted herself to a project long in mind: to get the family essentials packed in cases ready for instant evacuation in case of fire. At last she did it. To her surprise, Mother – although loudly proclaiming 'Hitler has his hands full in the Balkans' – did not object. The larger suitcase included all the lingerie collected between Munich and 3 September 1939, which she loved. A lighter cane hamper for Mother to carry. The bird, cash box, and deed box had priority, of course.

She went to a meeting and heard a talk on bird protection ('boring') that evening, home and completed diary by 8.30 p.m. ... to resume at 10 p.m. when the sirens sounded. They now had a bed and the sofa arranged in the living-room, with the bath and buckets full of water, windows wide open. Some 200 tons of HE fell around the town. She woke: '*Never have I gone through such a night of hell* – oh, no, not the bombs, they don't worry me, but the sofa. No springs and stuffed tight – my body is black and blue. Too narrow to turn on, too short, feet out in space and such difficulty in keeping the blankets from slipping off. What a night!' The single April attack increased Miss O's preparations. She prepared a hat box: two packets of birdseed, bandage, lint, burn cure, boracic ointment, scissors, two freshly tested gas masks, the memorial album to her late father. She also arranged three deep crown hats so that, in emergency, she could wear them all at once. Shoes were rejected as too heavy; from now on she planned 'to put on a new pair of shoes *whenever* the alert goes'.

Although Miss O faintly disapproved of excessive blitz talk in others, April entries bulge with her visits, often deliberate and devious, to bomb sites. She remarked (20 April) on the way she herself laughed at Miss B's office explanation of her chronic listening to Lord Haw Haw ('he comes on by accident') and wondered if she was 'like that with bomb craters ... I keep on coming on them by accident!'

At the end of April she was depressed by bad news of blitzed Plymouth (hearsay) and got into a bilious state ('generally do at this time of year'). 4 May there was a siren at 2 a.m.; gunfire only heard. They filled the bath, while neighbours abandoned all upper floors without fire-watching. Two beds now in the hall; cases all packed and ready to scram.

The next two nights came the last Clyde blitz. The diary stayed ominously silent. It remained so, exceptionally, until 12 May, when she suddenly resumed writing: 'The diary recovers full consciousness. I am sorry there has been a break, but I knew that my gastric condition was due to working harder than my strength can comfortably stand and last week I "lazed".' So ended (with over 600 planes) the last northern blitz of Clydeside.

A second diarist saw and heard it all from quite another angle, living over in the western isles, where they remotely experienced the glow of Clydeside on 13–14 March. Soon after, a college boy from the village was reported missing over there. His parents kept the village tea-shop in summer. Two uncles, herring fishermen, went over to help search the rubble. A third was summoned from his ship, in port. After an interval with no news, on 19 March, the diarist went across too. She did not get back home until 21 March, when she wrote up the trip as 'extremely moving' and impossible to write about at the time.

She arrived on the Clyde to find the bereaved parents there, in the flat of a friend, much distressed. They talked of immortality. The mother (Ellen) kept on asking if the lad had suffered, and told how on blitz-night she had dreamed awfully about her whipping him – so that she knew 'something had happened'. They discussed pity. Mother volunteered: 'I have forgiven the young chap that killed him – oh, long ago.' Father (P) added: 'He (the bomb-aimer) didn't know what he was doing.'

The three men were digging the rubble, relieving two more who were exhausted from the day before. The Demolition Squad had already given up, as the next house was likely to fall on to that side soon. The fishermen (D and S) just went on digging. The diarist stayed in the borrowed kitchen with the parents, holding a long discussion on heaven and hell, sin, guilt. Suddenly:

D and S come in; they were wearing blue overalls, half undone at the neck; but the overalls and their hands and faces were grimed with plaster; their hands were cut; S was looking wild, his hair on end and his eyes bloodshot; D's face was blotched with black shadows round his eyes and mouth and grey with plaster. *We got him*, S said, we got Jim, we got poor Jim, I told you we would get him . . . Then S began wildly: I have been drinking, he says I am a black sheep, I had a drink after getting poor Jim. I stood up and put my arms around his neck and kissed him and my mouth was full of grit and my nostrils were full of the smell of plaster.

There followed a verbal post-mortem on just how he died, with emphasis on instant death, repeated reassurance to Mother that he had not suffered. They were joined by the third digger, first engineer of a merchant vessel lately torpedoed in convoy – he preferred that shock to 'what he had just done'. He had been to report the find to the authorities. They would now have to identify the body. But first they had a hen S had somehow found somewhere, with tea and rhubarb tart and girdle scones.

Adventures followed. The diarist decided to stay the night with friends in a hamlet outside town. She got there by a train, a bus and a car driven by a drunk. She noticed the civic authority seemed to have lost control on that side. An army colonel had taken over in one place; he was feeding 1000 'refugees' from army kitchens: 'he'd cheerfully spent £8,000.' She got back to the group next day. 'They had been up and down having cups of tea all night.' They proceeded to plan the funeral.

Mother chose a 'white box for her wee boy'. She did not want a church burial, but hadn't liked 'to offend anyone'. No mourning though: so she bought a blue hat with a white ribbon. Cremation rejected because she wanted to lie by him herself, 'when her turn came'.

Father went out to steady up with a good plate of fish and chips and a Glasgow cake. S had disappeared – last seen with a tart (female, in trousers). He re-appeared at night 'with no hat and extremely drunk,' bringing a bottle of brandy for Mother, whisky for Father and making a general nuisance of himself too. They split up to sleep in two rooms. Our diarist shared a bed with Father, Mother, and two other young women, all in white flannel night-gowns ('only mine had no sleeves').

Eventually they made back for home, by sea, to wild reactions there:

Found a car, driven by their cousin and he took us off, driving fast. Sometimes Ellen said: I am being brave, aren't I? This is how Jim would like me to be. I said yes, and when we got in sight of Arran I quoted to her I to the hills will lift mine eyes, and she liked that and said He's there, isn't he? And I said yes, and here; time and place don't count for the free spirit. And she pointed out the wee cottage where her mother had been born; they were all crofters. Old Mac came to the door to see us by. She only glanced at the cemetery. As we came to the village her face was set; I came behind her and at once a wailing crowd of sisters in mourning burst into the passage and all over her. She began to cry, they were all crying: J in black, brought us a filthy drink, fizzy lemonade and gin – none of us had much to eat on the boat and it went slightly to my head and more to Ellen's. They all began to cry over me too, especially M. M began saying the most frightful things about poor wee Jim being too good for this world, he's happier where he is – all the *personal* things which I'd managed to get her away from. And how we are all going to die and it might be any day at all, and the old mother crying, and Ellen threw back her head and keened and howled and began calling for her wee boy. I went and held her hands hard and spoke loud and calmly to her till she bucked up; whenever I stopped for a moment one of the other sisters began. But I managed to get Ellen steady before P came back. He came straight to me and said I nearly broke down but not quite; I was brave. I said I knew he would be. Then I went into the kitchen; there was a meat meal for the men and I sat with them – I was very hungry. S began telling funny stories, mostly traditional stories about his boyhood. P smiled a bit. Then I helped to wash up; I had said to Ellen that now we were sisters. Everyone was running round, and I got a dry dish-cloth and wiped the cups and plates and brought them in again for the women. We had mutton and girdle scones and butter and cookies, and lots of tea!

(4) Not least, last – Hull

The keening cries from far Kintyre show how the blitz touched every corner of the islands. No place escaped the impact entirely. But the direct blows were on the few great cities and more numerous small ones. In human terms the smaller on the whole suffered most, bomb for bomb, per head of population.

In May the Germans started withdrawing their bombers for the prepared assault eastwards into Soviet Russia, openly launched on 20 June 1941. Well before that date, the blitz proper – the first great series of air attacks on civilians ever undertaken – was completed. Other, lesser series were to follow: the Baedeker raids on cathedral towns like Canterbury, the 'little blitz', the V–1 and V–2 unmanned weapons. But the deliberately indiscriminate assault on urban centres throughout the kingdom was never resumed. The latter attacks are important, too, only outside the scope of this present study. Putting them all into a crude perspective:

NUMBERS OF BRITISH CIVILIAN DEATHS FROM ENEMY ACTION

Year	*London*	*Elsewhere*	*Total*
1940	13,596	10,171	23,767
1941	6487	13,431	19,918
1942	27	3209	3236

Thus in 1940 deaths ran higher in London than outside, where in 1941 – overwhelmingly during the first half – the ratio is reversed, 1:2. In the total – 60,595 civilians dead from enemy weapons during this war – the London total was 815 below the provinces, largely because of the V–2 later on. Only 2% of fatalities were in Ireland.

Other towns and cities were blitzed. This writer and his colleagues were usually there in time or soon after on Tyneside, Cardiff, Sheffield, Nottingham, as well as Belfast; smaller attacks on Swansea (several), Sunderland, Barrow, Leeds. None of these, on the available evidence, shows any peculiarities of note. Only Hull, a town of 320,000 in the north-east, holds a special flavour. It may suitably round off this sequence of regional surveys. All the more suitably since Herbert Morrison considered 'the town that suffered most was Kingston-upon-Hull'. He suggested the Germans did not realize what they were doing there, mainly because the BBC, enraging the locals, referred merely to 'a north-east town'.[1]

Only 6000 of the 93,000 Hull homes escaped bomb damage by the end, from three main attacks in March and May plus many smaller raids favoured by an easy approach across the North Sea. Hull, as Morrison wrote twenty years later, was an 'easy target ... night after night Hull had no peace'. It thereby had solid citizen training.

An independent group of Hull citizens made their own study of the beginning of this extended experience. They reported how 26 'reception centres' dealt with 1773 admissions after a first but smaller (78-plane) March raid (17 dead). By the evening of 16 March, two days before the much larger (378-plane) blitz, 3294 persons were seeking help of some sort, 2216 of them re-housing. One working-class district had been evacuated owing to unexploded bombs. Many more were to follow.

Those local observers were particularly concerned about the relatively poorer St Pauls part of Hull, where they found morale 'low'. They arrived at the 'considered opinion' that St Pauls 'did not stand up to the raid shocks' as well as other districts. For once, M-O was not the only one to report this sort of way. The reaction in St Pauls was:

> One of complete helplessness and resignation. It was this attitude of resignation that provided the most disquieting feature. It was not a healthy willingness to accept misfortune without grumbling, but hopeless and indeed helpless incapacity to appreciate the significance of their plight, and the reasons for the disaster.
>
> Naturally we did observe some exceptions to this general condition, but they were in the minority.

The two gentlemen and one lady (city councillor) from the Hull committee went on, albeit a bit timidly, to suggest that in such districts particular problems needed more immediate attention, especially from the information services. As they put it: 'the general flow of Ministry of Information material leaves these people *untouched.*'

The very heavy raid of 18 March, a Tuesday, stepped up the pressure with nearly 400 bombers; 'incessant noise of aerial bombardment', from 9.15 p.m. to 4 a.m. To many it proved 'a nerve shattering ordeal', this time widely spread, in no way special to St Pauls. Yet Hull people picked themselves up quickly. 'The recovery was phenomenal.'

The 7–8 May double raid shook the place again. We estimated that around a third of the population had left the city by mid-month. Trekking was on a familiar scale. Raids continued across the North Sea into July, when the rest of the country was practically at peace again. An observer in the autumn of 1941 described Hull as 'the only town to have been heavily raided *since* the German attack on Russia'.

But a hard core weathered all storms to stay put, with steadily improving facilities but always a lack of deep shelters. This is how the landscape looked to a visitor after May:

The most specular ruins in Hull are along the banks of the river Hull (a small tributary of the Humber) – particularly the east side. Here one sees the still smouldering remains of the tall flour mills and stores. Ranks, the largest of all, Spillers, Gilboys, Rishworth, Ingleby and Lofthouse are completely destroyed. Only the CWS and Paul's remain undamaged. The workers from the ruined plants now work in these two. Further east and north-east, the industrial and working-class residential areas have suffered heavily. The huge Reckitts works (blue, starch, polish) are almost completely burnt out, and a large power-station is almost unrecognizable. The gasworks looks almost untouched, but Hull was without gas for six weeks after the May raids. Whole streets of working-class houses are down. One of the most impressive bomb-holes is on the Holderness Road. There, there is a crater almost 20 yards across, filled with greenish water, in which planks and barrels are floating. By the side of the crater stands the remainder of the Ritz cinema.

In the midst of such devastation a youngish artisan talked to his wife, reminiscing:

He: If it's got your number on it you get it. We've had it three times. Last time the house came down on top of us.

She: All the ceiling and the plaster. It was terrible. We didn't know where we were. And we couldn't strike a light ... we dursn't.

He: Mr Nicholls came round shouting 'Is there anyone dead in there? Is there anyone dead in there?' I said 'Let's get back into bed again and 'ave some sleep, we can't be any worse off than what we are.'

She: Then Mr Nicholls come and says the house next door's on fire, so we think we'd best get down the shelter.

He: So her ladyship says, if she's going down the shelter we've got to do ourselves up a bit.

She: We were covered in plaster – all in our hair and all over the place.

He: So she combs her hair, and wipes her face on a towel – and when she gets down the shelter she were black as a nigger – all the soot had fallen on the towel. (loud laughter)

She: And my hair was all in curlers. I put 'em in during raids, so as not to lose 'em. One time they got blown all over the garden. But we found 'em, didn't we? We picked 'em up, every one.

He: And after that we lived for six weeks in one room. We had to have the light on all the time – all the windows was boarded up.
She: And we'd no gas ... They wouldn't evacuate us. In the end, Tom had to *knock the house down himself* before they'd find us anywhere else. He pulled great holes in the walls with his hands.
He: It didn't need much pulling. You could put your fist through it.

Economically depressed by the loss of Scandinavian and Dutch trade the place just about staggered on. As late as May 1942 – when there was another heavy raid – the Research and Experiments Branch of Home Security remarked that 'Hull was, as it were, torpid and apathetic.'

Altogether, through the war, Hull weathered 70 large and small night attacks from piloted aircraft, as compared with Southampton 49, Bristol 51, London 251 (plus 101 by day). Life went on even for an old man, a warden, talking to a younger man while they went about the business of living through and after the blitz.

'I just went down to the Post an' when I come back it was as flat as this 'ere wharfside – there was just my 'ouse like – well, part of my 'ouse. My missus was just making me a cup of tea for when I come 'ome. She were in the passage between the kitchen and the wash-'ouse where it blowed 'er. She were burnt right up to 'er waist. 'er legs were just two cinders. And 'er face ... the only thing I could recognize 'er by was one of 'er boots ... I'd 'ave lost fifteen homes if I could 'ave kept my missus. We used to read together. I can't read mesen. She used to read to me like. We'd 'ave our arm-chairs one either side of the fire, and she read me bits out o' the paper. We 'ad a paper every evening. *Every* evening.'

(5) Around Worktown and the like: unblitzed

In October 1940, a housewife, aged 49, living in the south Lancashire cotton town of Burnley observed to herself:

This is a safe area on the outskirts of a cotton weaving town. No bombs have been dropped near here so far. We used to hear planes over here a month or two ago. Then the faint sound of distant sirens in a district towards Manchester. We were quite upset on first hearing them. Then we heard our own sirens a few days later. The first time was rather terrifying but on subsequent alarms we became more and more used to them, realizing there was little danger here. The last time they sounded was one early morning (2.15 a.m.) and I slept through.

She well illustrated the unblitzed aspect of blitzing. Until you lived through a really big raid, anything could cause anxiety. Familiarity, on the other hand, tended – among those who could take it and get over the first bangs – to breed a mixture of acceptance, pride and contempt. As she goes on to note 'air warfare has affected us very little but the *knowledge* that it goes on in other parts leads to a state of uncertainty and mental tension'.

Most of the more important towns between Merseyside and Manchester, around Manchester and down to Birmingham, were never blitzed or even noticeably bombed. Earlier in September, an Oldham (Lancs) observer barely mentions the possibility of raids but remarks that: 'A good number of above ground shelters are rapidly being built all over the town – but no one seems to take much notice and ask "Why are they being built?"' While in Wigan, at the end of George Orwell's pier, another wrote to say that, although German planes were heard passing overhead and searchlights seen to swing, interest in the war was 'falling off' to a deep low that autumn.

Set amongst Burnley, Oldham and Wigan lay 'Worktown', the name Mass-Observation gave Bolton when we studied it intensively, anthropologically, from 1937 on, taking it as a 'typical' industrial centre, population the size of Southampton's. Worktown passed through the war unscathed by direct German acts. In the locally peaceful September of 1941 a resident observer reviewed what appeared to have changed there in the first two years of world war. He noted a modest increase in drunkenness among juveniles and a drop in the number and availability of prostitutes ('many famous pros are in munitions'). People used the brick shelters through the winter on quite a large scale. Manchester, fifteen miles away, was too close to be psychologically comfortable. Most of the other marked changes were common to the whole country. Writing a year after London's blitz began, he could report:

Air-raids. Bolton has had one bomb. An eating house some 100 yards from Trinity Station was demolished. There were two deaths. People no longer go to the shelters except in very small numbers. The shelters were jolly places last winter, and Potato Pie parties were a regular feature in small public shelters.
Transport is very bad. There have been a great amount of extra buses for munition workers, but the general services are poor. The last buses leaving town at 10 p.m.
Dance Halls are crowded most evenings. I would not say there is an increase from normal times, but they are crowded.
Food. The situation is not bad in the poorer working-class neighbourhoods. Most homes where there are children have sufficient meat. I was in one home, and asking about the food problem generally when a nine-year-old piped, 'I could have my mum and dad locked up today if I wanted. I could have them under the clock. I've never had my bacon ration since war started.'

He went on to conclude that Worktown had developed an 'Inferiority Complex' – pet phrase of those Freud-clouded days. 'The people seem at once proud and ashamed of their solitary bomb,' he said.

The unbombed towns were nevertheless rich in bomb stories of a kind soon forgotten in the blitz, but well reflecting the pre-bomb anxiety. For instance, after one of the first warnings in Worktown, these remarks were overheard:

Young man: 'A policeman rushed out and it wasn't until he'd gone one and a quarter miles and spoke to someone he realised he hadn't got his teeth.'

Old man: 'A woman up Tuppers' Row lost her head and rushed out into the street with just a pair of stockings and shoes on. An air-raid warden said "Aren't you going to put anything else on?" so she went in and came out with a hat.'

Young woman: 'A woman ... came flying down Deane Rd in her pyjamas. While she was trying to do up her buttons her trousers came off.'

Young man: 'When the siren went I wandered round and looked out of the window and then went and had a shit. I wanted one, anyway.[1]

(6) Provincial Epilogue: back to London

It seems a long time since we left London in mid-November, to spend the winter and early spring provincially, ending with Hull in May 1941 and beyond. All this time, the metropolis went on being attacked, though with far less persistence than during those first two months. Still, over 5000 tons of HE were so directed after 14 November, or 21% of London's total. The later heavy raids were spaced without any discernible plan; 15 November (358 planes); 8, 27 and 29 December (the last a great 'fire raid'); 11 and 12 January; nothing much in February; 8–9, 15 and 19 March; 16 and 19 April (both *very* heavy, 685 and 712 planes) and finally 10 May (507). It was as if the Germans every now and again remembered their determination not to let London get too cosy, especially around the middle of the month.

While London was never exactly peaceful, it had lived out the worst of its unease by November (Chapter 5). Most kept up the patterns learned all through winter into the spring, adapting these personally so that, say, they slept quiet nights upstairs in the house but went downstairs or into a shelter when it seemed desirable – a rather more elastic pattern but no longer improvised. If the Luftwaffe had ignored London for, say, three months, that pattern would perhaps have been broken; but they did not do so. Diarists spelt out the story. Some barely mentioned raids in the New Year, though anything 'different', sudden, peculiar could readily disturb the calm.

One example: a single large bomb exploded with great effect in Hendon on the night of 13 February, in the same London district as some of the earliest bombs. More than a week after this incident, locals were still speculating on what it was: a land mine *without* the usual parachute; an aerial torpedo; several bombs tied together; an outsizer. What worried most was the 'different' noise it made while falling. Two men, then three women, indicate the anxiety always underlying calm human surfaces – providing opportunities for the sort of upset which the Germans never followed up during the blitz.[1]

'It didn't sound like a bomb falling. It looks like a landmine the damage it's done and the row it made when it exploded. It was an incredible noise.'

'If I'd thought it possible, I'd have said it was a shell – it seemed such a long time coming and it seemed to have that spinning motion. But as it was I think it could only have been a large bomb dropped from a great height.'

'It definitely wasn't an ordinary sort of bomb. It didn't *sound* like a bomb. I remember turning round to my hubby and saying: "That train sounds close" then the explosion came and gave the house such a shaking.'

'I know what it was. An aerial torpedo. It was the noise it made coming down. I couldn't make it out at first but afterwards I realized it was the propellors making the noise.'

'One bomb, however large, couldn't have done all this damage. And I know by the noise it made when it came down. I could distinctly *hear* about 7 bombs *and the chains*.'

But when another small bomb soon fell close to the puzzler, no one bothered much. 'I call this piffling little thing an anti-climax.'

A fortnight before that mystery package upon Hendon, a more serious break in routine had occurred, 29 December 1940: 'The Great Fire Raid', largely affecting the City. Under 150 bombers were responsible, with around 600 incendiary canisters. This was much less than Coventry, and London itself had nearly double (1150 canisters) on the day after that, 15 November, then over 3000 on 8 December; more than 12,000 canisters altogether had been dropped in the opening September–November London series, though with only two nights then above 600 (to 718 on 10 October). Never again, however, was the effect so great. It was the Southampton story again: a Sunday night (Christmas holiday week): shop and office blocks closed up, churches locked, most of the fire-watchers off duty. For just as the fire services were still fragmented and months off being 'nationalized', so fire-watching was still on a voluntary basis – although ARP and nearly everything else of that kind had been regularized on to a compulsory service basis long before.

This probably was the Luftwaffe's one *clear* victory over London, its effects fully commensurate with the cost and effort. A strong westerly wind and low tide helped. Thames water could not be pumped and the ordinary mains went almost dry. Little more than two hours of attack set going some 1500 fires, some in the East End but 90% of them in the heart of the City. Unchecked, in locked buildings, and soon without water, most firemen could only watch as 'the biggest area of war devastation in all Britain' burst in a great golden sun-flower. Leonard Mosley has summed up: 'The tragedy was this: for more than four months every householder in London had been dealing with incendiary bombs and had lost his fear of them.' But this is an incomplete explanation. Less fear need not mean less attention; on the contrary, people continued boldly fire-watching *for themselves*. More and more, however, as the war continued, they came to feel that extra duties, outside the family, should be allocated by government on a national, equitable basis – like joining the army or other 'public' service. On the eve of 1941 Herbert Morrison, Minister of Home Security, broadcast over the BBC a speech which clearly blamed the public for 29 December. He introduced compulsory fire-watching by registering all males between 60 and 16 for up to 48 hours' service each month. On that December night, over half the fires could have been avoided if some such system had been earlier enforced. The Labour Mayor of Stepney wrote that 'with proper fire-watching precautions it need never have occurred at all'. Although the Germans tried to repeat the dose, it did not work again. For as the Mayor put it: 'Now we are ready – after the event.'[2]

This strengthening of the fire-watching system also had a more general effect, lasting after the attack on the last Saturday 10 May – as noticed for instance by a West London commercial traveller who came east on Monday 12 May and wrote in his diary:

During business calls meet fire watchers who had their first 'blooding' in Saturday's heavy raid. The spirit of the people seems to be moving from passive to active and rather than cower in shelters they prefer to be up and doing. Incendiaries seem to be tackled as though they were fireworks and tackling fires in top rooms with stirrup pumps is just part of the evening's work. One leader was telling me his chief trouble is to prevent people taking risks. Everyone wants to 'bag a bomb.'

Over to the south-west of London the park-keeper whom we last saw digging a hole for his Anderson on 3 September 1939, by now took little notice of anything except the incendiary threat. On the last big night of 10 May (over 500 planes, 700 tons HE, 7000 cans of incendiary) with high moonlight in a cloudless sky he went to his shelter at 11 p.m.

I remained in the shelter looking out once when I heard a shout outside. I saw nothing however ... I lay still waiting for the noise of guns to cease so that I could drop off to sleep. Once a plane dropped three bombs, near enough to rock the shelter. No further bombs fell locally. My chief anxiety was over incendiary bombs. I saw none nor any signs of them. Sleep was absolutely impossible. The drone and roar of the planes went on all night. I dozed however and was wakened by the all-clear signal just before 6 a.m. This was time for me to get up. So I got up feeling rather tired.

Next morning, Sunday, some complained of lack of sleep ('I had no sympathy for them'), while some older people (65–70) said the renewed attacks were wearing them out. For this sudden ferocity was hard to take. Fires, over 2000 of them, though not concentrated on any centre, still spread widely, with the Thames again at low ebb. A new record for dead was set: 1436, with another 1792 seriously injured. The Tower, Law Courts and Westminster Abbey were hit.

More of the same might have shaken Londoners savagely. But a well-off widow, 28, two children, in W.9, after a disturbed night during which she tried, in vain, to reach her duty post at a centre, was still able to write in her diary next day:

I drew back the curtains on a day of sunny loveliness and perfect peace. The apple-trees in the garden were pink-dotted against the luscious, thick-piled whiteness of the pear-blossom; the sky was warmly blue, birds were chirruping in the trees, and there was a gentle Sunday-morning quietness over everything. Impossible to believe that last night, from this same window, everything should have been savagely red with fire-glow and smoke, and deafening with an inferno of noise. Even when one noticed the gaping patches of jagged blackness in the sun-brightness of the windows opposite, and the ever-present sound background of the fire-pumps humming drone from down the road – even when one came to realise, disgustedly, that the water was running thinly and in fits and starts only, and that the gas was as dead as a doornail (of course it would have to be hit in a main *just* when, after a deal of trouble, we'd at last managed to get the year's last lovely joint of pork for a Sunday roast!) – the events of the previous night seemed as incredible as the fantasies of a nightmare.

She sat back to take stock of her nerves:

I was rather disgusted with myself for feeling so feeble and exhausted all day, as I didn't have nearly such a tiring night as most Londoners must have undergone. But on the other hand it was a very great relief to me to find that after all my anxieties of the last months my blitz-nerve, when it came to the point, proved after all not to have 'gone' during the prolonged lull – I hadn't been any more frightened (for what that's worth) than in the old days when blitz had been a commonplace of life.

The park-keeper's lack of sympathy for unadjusted sleepers and the widow's relief that her own adaptation had lasted were echoed another way in about the best of the published blitz diaries, that of the U S Military Attaché, General Raymond E. Lee, who soldiered through the whole London blitz. On the night of 10 May he was at Claridges Hotel, W.1. Soon bombs shook it 'from top to bottom'. He went out to the balcony to look, then read in the lobby where he was joined by noted American journalist Vincent Sheean (*Personal History*) just in from the States. Sheean and his wife were nervous, 'upset at every explosion'. The General was bored by the conversation and took the newcomers on to the roof, to see fires all about. There was 'the usual drone of the falling bomb', a huge explosion. Then he heard 'a sudden convulsive movement' and looking round, found 'both Sheean and his wife grovelling flat on the tin roof, which seemed to me rather shocking'. After a while they got up. They hardly seemed to realize that 'after a bomb has gone off there is not much use falling down'.[3]

This does not mean that all the old hands were steady, of course; and especially that, even with incendiary bombs, they were 'unselfish'. An observer in W.C.1 on 10 May complained bitterly of neighbourhood attitudes in Ridgmount Gardens, where residents 'showed a marked reluctance to take any action likely to endanger their own skins'. Great difficulty was experienced in getting anyone to help deal with incendiaries on the roof. 'Although they were repeatedly told that the roof was alight above them, practically none of the residents huddled in groups on the first and ground floors would come upstairs and help, and those that did things soon went away.' Just as in Glasgow – and some of the time everywhere, too.

Some things went on unchanging from the first bombs on London to the last. Conspicuous always was the desire to see new damage. An observer watched the familiar ritual outside Westminster Abbey, then reported the reactions of local residents and officials, on the following Monday (12 May):

'They came round here all yesterday, getting in the wardens' way, with their cars, sightseeing out of morbid curiosity. To think there should be such people.'

'All the sightseers yesterday cleared out all the cafes and things round here so that as the ARP people came off duty they couldn't get anything at all to eat or drink.'

'All those morbid people here yesterday. They ought to be shot, every one of them.'

The standard for debris-watching had gone up over the months however, so there was some disappointment.

'It's not so bad here as they made it look in the photographs.'

'What's happened to the Abbey? It doesn't seem to be *touched*.'

A rarer reaction was: 'All these lovely buildings destroyed. I *hate* the Germans.' There was a sanction for West End sightseeing too: 'Churchill's coming round to see it today. He wants to see for himself.' He generally did.

Further north, another observer found less interest in the huge mess at King's Cross station, with a large gap in the glass roof. 'Half a dozen soldiers shovelling rubble into a lorry and a dozen or more sat or stood around.' The 'great majority' of travellers on the platform took little or no visible notice of the damage. Similarly at St Pancras Station, closed because of a large crater outside, 'the passers-by only gave a casual look as they went'. Londoners were getting used to almost everything...

In May M-O's national panel reported their feelings about what was going on in London, with special emphasis on the blitz. Our monthly bulletin for that month compared three groups of reactions – from resident Londoners, evacuated Londoners and the rest looking on or into London – and concluded:

The resident Londoner, in contrast to the evacuated Londoner, appears least affected of any group by what is going on. Usually he (or she) is glad to be there. Unlike evacuees, journalists and people living in more remote provincial areas, he seldom eulogizes Londoners. A stretcher bearer in N.W.1 writes:

'When I pass through the streets I see them looking much as they did in peacetime. That cheers me. I am glad of such entertainments as there are . . . The other principal thing that is apt to happen in London is the bombing. I resent that. I resent the fact of people being exposed to it and its futility and cruelty. My feeling about London is one of reasonable content tempered by *boredom*, though I am often quite happy, especially in the fine Spring.

A woman ambulance driver in S.E. London:

I have my favourite places I hope will not get hit – for the rest it's just part of the war that's got to be put up with. I feel much the same as I always felt – passionate affection for it – a feeling of release and freedom. Feelings unchanged but maybe stronger.

In many Londoners' replies there is a simple view. Woman, 27:

I wouldn't live anywhere else. I have grown *fonder of the place since the blitz began*, and I find it quite exciting to spend Sunday afternoon in the centre of London, Charing Cross and Leicester Square.

Evacuated Londoners often have intense feelings about London itself and about the bombing. A longing to get back is common. For an evacuated girl of 18 from the East End this is a possibility:

I am longing to get back to London after June when I have taken Higher Schools, to be in it, because I feel that victory depends to a large extent on London, which is as it should be.

We must not leave London with the impression that all was perfect there compared with the wretched provinces. It was not so. The presentation of this study, in rough sequence, may have tended to give a little more emphasis in that direction than is really justified. It is difficult, now, to be sure. But clearly Londoners did enjoy enormous advantages, in places to go to (sideways and down) *within* the broad community, in leadership and in services, especially because of the quick access to those of central government right into No. 10 Downing Street. Churchill needed to devote only twenty minutes of 12 May to inspect the Abbey debris.

For a very last item of on-the-spot observation, glance for a few moments at the main Paddington Rest Centre, now well equipped, on the May Monday at 5.30 p.m. A few children, looking 'thin and unhealthy but perfectly cheerful' play outside. Inside 40 people, mainly under 15 or over 50 and female, eat tea. 258 had officially registered as homeless inmates, but at 2 a.m. that morning a bed count showed only 89. The rest had mostly gone Underground. The Centre Superintendent, 40, said people were more subdued than last autumn. The Matron thought many were 'just tired out . . . simply worn out. Can you wonder?'

The Super added that this raid has taken Paddington 'completely by surprise'. Not *one* doctor was available at the weekend. The billeting authorities were down to skeleton staff. And so on. Oddly enough, *The Times* that 10 May morning had a double column leading article, ready for the fiery dawn: FIRE AND AFTER. It is packed with belated criticism of fire-watching and fire fighting generally, moving on to rescue parties, Rest Centres, lack of co-ordination, lack of Advice Bureaux and information, Emergency Committees. It 'cannot be left to the chance of whether the EC of the local authority happens to be complacent or energetic, obscurantist or forthcoming, prejudiced or imaginative'. But that's just how it was left as the Luftwaffe left Britain for the Russian front – except that those six adjectives are insufficient. One could add five more, all frequently employed: SENILE or OLD-FASHIONED, UNINFORMED, MISLED or STUPID.

Field-Marshal Hugo Sperrle threw in everything the Luftwaffe could spare for that last fling before the bombers had to turn eastward. Unflinchingly he clung to the classic Douhet thesis, so late in the night: 'There is no foe that bombing cannot break.' At the receiving end, Churchill later wrote of this: 'the worst attack was the last.' Coming as it did, the climax staggered some Londoners afresh. They swayed in the blast – but little more. What five more nights at that intensity *might* have done is another, almost academic, question.

9 Morale Questions

'We Vietnamese are used to the war,' said my interpreter. 'We have suffered so much that we can laugh and cry at the same time.'

JON SWAIN
(*Sunday Times*, 6.4.75)

What did all this bombing of Britain achieve – other than to destroy buildings and dislocate and distress millions of humans? Did it look like helping to win the war for Germany? And in particular, did the effort and expense in men, machinery and raw materials justify the results as compared with what might have been achieved by other means?

It is this writer's laboriously considered conclusion that, *as it was done by the Luftwaffe*, the answer to the first two questions is: 'very little', except in a transitory way. The answer to the third question is surely 'No', and, incidentally, it probably applies to Britain's air attacks on Germany almost as well. But this leaves open a subsidiary question: had the methods been directed more closely by accurate intelligence and patterned so as to achieve the maximum impact, could the results have been far greater? And here one suspects the answer just *might* have been 'Yes'. In particular this could be so if so much of the initial effort had not been directed at London; had, say, Coventry been blitzed first and London for not more than three nights at a time at first? Or there are any other possibilities, including the support of crude bombing by other means and media.

The blitz was a terrible experience for millions, yes. But not terrible enough to disrupt the basic decency, loyalty (e.g. family ties), morality and optimism of the vast majority. It was supposed to destroy 'mass morale'. Whatever it did destroy, it failed over any period of more than days appreciably to diminish the human will, or at least the capacity to endure.

This raises the further, perhaps the final question. Can blitzing *ever be terrible enough* to shatter the 'morale' of an ordered society which feels basic confidence in its own structure, fairness, inner organization and good faith – even when some of these have been fleetingly obliterated? The answer to this one goes to the roots of strategic bombing policies in the Second World War and since, notably in the Vietnam war. It partly depends, in the end, on what you *mean* by morale, what morale 'is'. For the whole concept behind these huge air efforts was that civilian morale could be 'broken' regardless of the armed forces, the 'nation' thus brought to its feet by overleaping the military, as enunciated by General Douhet in the twenties. To get nearer those twisted roots, we must first re-examine morale in general, as it was seen by those who planned to break it. We must also sum up the general effects of bombing as measured in more specific behaviour, in public opinion and in memory, short-term and long. We can then, in conclusion, return to the basic question.

Under all the varied circumstances, the final achievement of so many Britons was enormous enough. Maybe monumental is not putting it too high. They did not let their soldiers or leaders down. Not infrequently, indeed, they propped their leaders up, in a situation where leadership at the local level was lacking. In official and much other thinking the quality primarily identified with this achievement was called 'morale'. The noun has been used in earlier chapters, as the only term of the time. It was borrowed from an earlier usage, primarily for uniformed troops, as in the pre-war Oxford Dictionary. There is even a hilarious account of how the Austrian Ministry of the Interior tried to measure 'the morale of the members of the local government and intelligentsia', in the classic satire of the First World War *The Good Soldier Svejk*. They invented no less than twelve grades for rating 'unshakeable loyalty to the monarchy'. No one did anything so refined this side of the channel in the Second World War.[1]

Looking back, one can well see that this term (as we used it then, applied to the whole civil population) concealed certain serious confusions due to lack of thought. By the Plymouth blitz – as the reader may recall – we found it necessary to distinguish at least five main levels of morale. Long before that, several of us in this field had been worrying lest we were really muddling up all kinds of human qualities inside one convenient six-letter word; and, by so doing, complicating, perhaps compounding the real problems rather than clarifying them.

When Herbert Morrison lamented the loss of Plymouth's morale, he was voicing not only the current cabinet concept but also a deeper confusion. As indicated in the discussion of 'defeats' and elaborated by later observation, many people were more or less depressed and much distressed, by provincial blitzing especially. Many quit undamaged homes to avoid more of the same. In doing so, however, they did not go berserk, shout 'Fuck Churchill!', loot the unwelcoming villages or rape the clergyman's wife. They required neither priest nor psychiatrist. They had to make do mostly for themselves. They did so. And they continued working as nearly as possible as before, in added difficulties of transport.

Morrison, in the contemporary Whitehall mood, was identifying morale primarily with *cheerfulness*. The stereotype of a good patriotic citizen was supposed to laugh, to make the V-sign when he saw a cabinet minister, and so on. A lot of them didn't do much cheering even at the best of times. But they could cry and carry on, all the same.

'Morale' proved such a mishmash of fantasy and fact that it could vary almost in direct contrast to outside events, so that a real bomb could be 'better' for morale than the dread of one in dreams. Some people panicked at first for almost *nothing* – for the first siren set up by a French plane gone astray that September Sabbath in 1939. Some of the same people, as our diarists show, most fearful at first, later stayed in bed through full-scale blitzes.

Man's day-to-day, erratic resilience was the vital factor ignored in pre-war prophecies. Looking at this more closely for a little, at one stage we had some one hundred mass-observers plotting their daily and weekly changes of mood in war, marking themselves against a personal index. In the same week, the men rated themselves overwhelmingly cheerful or pleased ('good morale') with a peak of 203 on Saturday, 188 Sunday, a low of 110 in the week. Women showed much readier depression, as usual, with a Friday low of 56 and a Saturday high of 120. That was one reason why weekend blitzes, like Southampton's, were more of a surprise to this resilience. For many Britons the fun of living centred on the weekend.

M-O made many other efforts to analyse morale factors once war began. We should have started in 1937 at the latest. The advent of war clouded the concept with issues like loyalty, patriotism, unity, treason. For a while, we asked the national panel of voluntary observers, urban and rural, to keep a daily chart on themselves, in the middle of the blitz, December 1940, logging 'the main things ... which determine day to day feelings ... in fact morale'. 211 people completed all their charts. For once there was a high degree of consistency in results, not a complicated criss-cross. Simplifying the broader conclusions, these were the main influences regarded as keeping up one's spirits, in order of frequency:*

Factor	*Men*	*Women*
Friends	–	–
Health	2	2
Sleep	4	3
Work	3	5
Weather	8	4
Meals	7	6
Recreation	5	8
Love life	9	8
Weekend	6	12
War news (if good)	13	7
Introspective thoughts	10	10
Family	11	11
Money problems	12	13
Non-war news	14	14
Journeys	16	15
Air-raids	15	16

These same people asked to rate what *depressed* them most, put weather first (especially women), then general war news, well above air raids.[2]

*The figures in this table have been questioned but since Professor Harrisson's source could not be traced it was thought best to leave it as it stood.

Rating lists of this sort hardly begin to indicate the intricate balance of pros and cons, on a morale barometer. That may be better seen, perhaps, if one takes the total votes for each factor – that is, the number of references to that factor without regard to its supposed effect. This puts air-raids in another perspective.

Factor	*Percentage of both sexes voting this as important*
1. Health	44
2. Work	39
3. Weather	38
War news	38
4. Friends	37
5. Introspective thoughts	36
6. Sleep	35
7. Family	30
8. Recreation	27
9. Meals	26
Money problems	26
10. Love life	25
Air raids	25

This admittedly very rough echo of what people felt, before the start of 1941, does indicate how much the constant preoccupations of peace persisted. Ordinary 'news' would probably have been placed quite high in peace too. In any case air-raids of course affected almost every other factor. Nevertheless, their placing is low compared with the emphasis necessarily given to them in this study. We may find reason to confirm this view, in some respects, when further examining the long-term effects of the blitz.

We also conducted, every Monday and Thursday throughout the war, small sample surveys among adults in London and the provinces. On this was based what we called a 'news index': the amount of interest being shown in war events. Correlated with this went an analysis of 'cheerfulness' and 'depression' about the course of the war, as expressed verbally by men and women in the street. One of several significant trends suggests the apathy which never really vanished and which actually came to a peak in the blitz, at over a third of the whole urban population:

Month		*Average per cent per month not following the war news*
1940	May and June (before 'Battle of Britain')	9
	September (blitz begins)	14
	October (full blitz on London)	22
	November (Coventry, etc.)	29
	December (Greek successes)	26
1941	January (African success)	34
	February (African success)	39
	March (African success)	36
	April (African and Greek defeats)	33

Crude though it was this index showed a remarkable degree of detachment, which tended to rise with blitzing, when people became more and more preoccupied at a nearer, more personal level. In towns like Portsmouth disinterest rose to *over 90%* in some phases of the blitz (spring 1941). We earlier saw how often also they saw the war as nearly over, there.

This serves to emphasize that, up to the point of stress reached in blitzed Britain, people could carry on with their own (adjusted) interests, without taking any lively interest in important outside events or indulging in fun and laughter, 'high spirits'. It took a long time for the official leadership to start to recognize this. Those concerned with guiding the public mood, mostly through the Ministry of Information, for long took a contrary view. They urged YOUR COURAGE, YOUR CHEERFULNESS, YOUR RESOLUTION, WILL BRING US VICTORY, in a poster which exactly defined the distinction between the Minister (we) and the mass (you). A major campaign sought to seal the safety valve of 'idle talk', rumour. Although such rumours could go far, as with martial law on Merseyside, the supposed enemy agent would have had a hard life making use of such stuff. Rumour also defused anxieties; its rich persistence, heedless of official propaganda, helped distract attention from local grievances, which were usually *less* after blitzing than before – thus saving the authorities from much and sometimes justified abuse.

Churchill cancelled the rumour campaign. And by appointing his friend Brendan Bracken as fourth and last Minister of Information, after the blitz, he belatedly imposed (July 1941) his own concept of 'resolution' upon government. Bracken had long since beaten his own curious track into public school at Sedbergh, where young males in their formative years were expensively introduced for a few years to the virtues of unnecessary hardship (including hunger) and – until attaining prefectorial rank – artificially induced group anxieties. Lord Ismay, Churchill's Chief of Staff and intimate through 1940–45 well said of his old Harrovian boss afterwards: 'He was inconsiderate. Well of course he was – to all of us – but he was far more inconsiderate to himself. The only thing he thought of was the war. Whether anyone was made uncomfortable or whether he was made extremely uncomfortable, didn't matter a damn – all that mattered was to win the war.' The less expensively educated took these virtues as the norm: from infancy to death. Putting up with discomfort, enduring economic uncertainty and periods of familiar distress, were built into pre-war 'working class' experience. If the façade of acceptance cracked occasionally, it was only occasionally: the Hunger March of the early thirties was not so very different from trekking in the early forties. American columnist Dorothy Thompson once, early on in the difficult days, asked Churchill:

D.T.: 'How are you going to win the war?'
W.C.: 'First we shall see we do not lose it.'[3]

Many, not least he, had feared civilians under pressure would do just that, would lose on their own battlefield. They did not do so. They merely staggered, retreated, re-grouped, re-arranged themselves. Under the circumstances, this was their major contribution. They could contribute little more than to muddle through, obey the military law and maybe mutter a little. This fragmented achievement was won at the cost of extreme and often unnecessary discomfort, immense to the outer observer but not necessarily so deeply felt by most of those living with and through it. Seen the other way round, this was just another link in the long chain of the evolutionary adaptation of modern man, incessantly, out of the Stone Age, from Eskimo to Astronaut. The famed American general's remark on bombing the North Vietnamese 'back into the Stone Age' missed the point. The US were bringing another people, with a bang, into the new airborne civilization of napalm and lethal nails, closely supported by plastic and porn from the ground forces.

But before the blitz began, nothing had happened to make those at the top take a brighter view of the resolution of those beneath them. Churchill proudly described 'The Dunkirk Deliverance'. Physically it was a deliverance; psychologically, Britain as a whole got nearer private panic than ever before (or since?). Just before that 'miracle' in the spring of 1940, the Premier received a paper signed by all three Chiefs of Staff and their Vice-Chiefs, which in thirteen paragraphs, summed up their thoughts for the year(s) ahead. Of these the most relevant are the tenth and the last:

> 10. If therefore the enemy presses home night attacks on our aircraft factories (Coventry and Birmingham especially mentioned) he is as likely to achieve such material and *moral* damage ... as to *bring all work to a standstill.*
>
> 13. Germany has most of the cards, but the real test is whether the *morale* of our fighting men and civilian population will counterbalance the numerical and material advantages which Germany enjoys.[4]

Well, as we now know, whatever happened to morals and morale, work never came to a standstill for more than a matter of hours in any industrial area, Coventry and Birmingham included. But the Chiefs of Staff did bring the whole issue right into the cabinet. From this time on, high level interest in preserving and improving morale, in stopping moral sickness, increased and became more particular. Those few of us directly concerned with the study of the masses *en masse* found ourselves more and more engaged in assessments of morale. This in turn led, quite soon, to increasing uncertainty regarding our terms of reference, terminology and the like. What were we trying to measure? And how were we to do it?[5]

Imagine a doctor – or even a psychiatrist – trying to diagnose a disease that had a name but no defined symptoms! This was our problem – and a ripe source of confusion. Among various measures, several committees were developed to sort out what quickly began to look like an intellectual mess. To reconstruct the sequence of events is to be about as obscure as the subject of morale itself. Three distinct groups primarily concerned with the subject can, however, be identified in our own papers. One committee I sat on myself and can vouch for, although none of its documents seem to have anywhere else survived. Without going into detail, the overlap between the three groups can be indicated briefly by quoting parts of two letters written in 1973 by distinguished civilians, who played leading parts in the field of propaganda and the raising of morale and were both directly involved in this elementary attempt to analyse the problem in order to act effectively.

First, my old friend, and leading Labour politician, the late R. H. S. Crossman, who later on was to direct political warfare against the morale of Germany.

My Dear Tom,

Yes, I was on the Morale Committee of the MOI for some six months before the outbreak of war and then for a few weeks after war was declared. It was an utterly futile committee dominated by the formidable Lady Grigg (wife of P. J. Grigg who was later Secretary of State for War.) Among its members were the drunken . . . , one Gervaise Huxley who in peace time was in the tea market, and myself. None of us knew why we were there or what we had to do and soon after the war began we packed it in.

Yours

Dick Crossman

Second, Kenneth (now Lord) Clark, who as Controller of Home Publicity in the Ministry of Information during the blitz, sat with a high level group there, whose activities are authenticated by minor documentation in the Public Records Office.

Dear Tom Harrisson,

I kept no papers of the Ministry of Information times as the whole episode was distasteful to me and I was conscious that we achieved absolutely nothing. This is particularly true of the Committee you refer to, of which the full title was The Home Morale Advisory Committee. I was a member, and I believe later Chairman of this Committee. I know that Brendan Bracken (before he became Minister) bounced the old Canadian Prime Minister, Lord Bennett, on to the Committee, and I suppose that this senior statesman must have originally been the Chairman, but he was so slow and ponderous that nothing happened. The only other member of the Committee whose name I can be sure of was Mollie Hamilton.

In retrospect the only interesting feature was the amount of evidence that came in on how low morale in England was, much lower than anybody had ever dared to say. But there was obviously nothing that we could do about it.

The third (my) committee, called bluntly 'The Morale Committee', included the late Sir Julian Huxley, FRS, the royal doctor Lord Horder, journalist Francis Williams, psychoanalyst Dr Edward Glover and the Editor of *Picture Post*, Tom Hopkinson – who cannot remember anything about it. We had long, anxious meetings, until Huxley fell out with the forceful Director-General of the MOI, Frank Pick, and we disintegrated. The Director of Home Intelligence, Mrs Mary Adams, had formed this group and continued to concern herself deeply with the problem far into 1941.

The seeds of disintegration were planted in the very concept of morale. Morale could hardly bear close analysis. No one then around could come up with clear definitions, let alone clear programmes for dealing with the matter once it had been defined. So any ills could be attributed to loss of morale, thus by-passing the real human issues. If things went badly, a supposed 'break' in morale could be blamed without facing up to the background factors which caused the break: resentment at patent unfairness, confusion, misunderstanding, cold weather, lack of entertainments, inadequate leadership, utter boredom, for example. When Morrison at the cabinet committee on the Home Front, saw Plymouth's morale as 'gone', he in effect ignored the prime consideration: the failure of government at most levels and especially of his own Ministry to react quickly and competently.

Alas, all this is not a mere matter of semantics, of academic distinctions. The failure to analyse and define the complex of moods covered by 'morale' led to major deficiencies, both in writing about it accurately and dealing with it usefully. As regards the latter, Lord Clark's last sentence is to the point: propaganda probably could do very little indeed. But facts, hard information, could often have been supplied by his Ministry for one, with great effect – not directly on morale but contributing to a reduction of distress. A large part of the distress arising after a blitz derived from the lack of information.[6]

We seem to have given up the word-game in 1942. By then the

war was running more Britain's way, so that concern for defeat, Douhet style, was much diminished this side of the channel. The confusion of thought was now related to attacks on Germany. After the war, the term continued in its old, wild use. There is an enormous post-war study, *The Effects of Strategic Bombing on German Morale*, published by the U S Government Printing Office, Washington, in 1947, which typically treats morale as something going up and down with bombing, a sort of invisible cousin to medical shock. As the best of American social-psychology critics, Dr Irving Janis of Yale, pointed out (in 1951), the reality of the supposed relationship between bombing and German morale was never demonstrated in a scientifically or quantitatively acceptable fashion. After incessant reading this writer can only concur: the huge US study is, in the last diagnosis, a verbal mess.[7]

Today the word 'morale' continues in common misuse by press, presidents, generals who probably never heard of Douhet. It was South Vietnam's morale which collapsed, says this morning's radio, thus blocking out a whole saga of ecology, sociology, economy, politics, money and corruption. Nor was this true of the Anglo-American language alone. It was and is so everywhere. There is ample evidence that the Germans bombarded their wretched secret agents in Britain with demands for reports on morale of the same sort as M-O was struggling to provide for Home and for Naval Intelligence. (As those agents were nearly 100% controlled by MI5 after Dunkirk, they fed cheering estimates to their masters on the continent – in the way described in Sir John Masterman's delicious *The Double Cross System* [1972].)

The failure to define morale proved as important in air-raid planning and ARP as misdefining the physical and psychological effects in ways discussed at this book's beginning. The target aimed at – whether by pilots or propagandists – was vague and elusive. Any chances of hitting it successfully were thereby vastly reduced. The bombs scattered about Europe in the Second World War were at first aimed at armed positions and troops, then increasingly at military and industrial targets, until the blitz initiated random distribution over any dense agglomeration of humans with their allegedly vulnerable morale. This morale-busting had, willy nilly, to be directed at an invisible and amorphous lump. No attempt was made to distinguish what parts of it needed special treatment, or how the malfunctions created by bombing could be extended and perpetuated.

How then bomb millions of minds, souls, whatever, back into the stone age, forward to defeat, or anywhere else? In early 1943 Dr C. W. Emmans of Home Security wrote an important paper around this point. 'Morale is a word without scientific meaning,' he firmly declared, going on to suggest new ideas for comparing British and German raid effects, with disturbing implications for the simplicist view carried over from General Douhet, and persisted in by the RAF and Air Marshal Harris, Cherwell and Churchill, even after the Luftwaffe had amply demonstrated some of the method's defects. Simplicism won, as it naturally does if unchecked by exacting analysis. It still prevails. It dominated US policy in the whole Indo-China war. In a sense, therefore, all these decades of bombing were aimed at nothing.

10 The Learning Puzzle

PROVINCIAL ... having the speech or manners or *narrow views* practised in provinces.

Pocket Oxford Dictionary, 5th Ed. 1969.

'No REGIONAL body must now presume to treat Birmingham as if it were a hamlet of Nether Backwash.'

NORMAN TIPTAFT
Chairman, Birmingham Civil Defence Committee (in the blitz).[1]

The main objectives in ARP planning were the minimization and control of destruction to property and death to individuals, search and succour for the wounded and such measures as could be contrived to treat shattered nerves and prevent mass panic. Accepting the initial breakdowns as in part resulting from faulty diagnosis, as well of course as from the unprecedented character of the bombardments themselves, what emerges from field study is, most clearly, the way in which the same mistakes continued to be made, often without signs of any serious effort to correct them. The consequent additional burden of distress was tremendous; and, in theory at least, largely avoidable. What happened? Or, more exactly, what didn't happen?

Anyone travelling the long route from Stepney through Coventry, Southampton, Portsmouth, Bristol, Clydebank, Hull, could hardly have failed to see – though he might not dare to report – the same sad things happening again and again. A serious recognizable mistake might be observed without much surprise twice; but why twenty times? London demonstrated what could be done to correct errors in preparatory thought. That was not so difficult, given its huge resources. Merseyside was rich and great also, yet apparently had not heard word of London's experiences.

M-O's 46th blitz report, officially filed (and paid for by) government, stated in May 1941: 'We have repeatedly shown the breakdown of local and national measures which are supposed to provide for human welfare. No unit conducting objective field-work could have avoided seeing the inadequacy, the pre-blitz mentality of some of the measures provided. Yet it is impossible to see that anything serious has been done.' (FR 683.) We were not alone in voicing these views. We have already heard two key civic leaders, Willink of London and Astor of Plymouth, declare that the provinces never benefited from London's early training or from each other's experiences. Lord Astor went further in a confidential paper; he considered that not only did local authorities 'not profit from each other's experience' but moreover that both Regional HQs and Whitehall failed to give needed advice and leadership. Such powerful criticism, rare in wartime, fell far short of giving the whole picture, however. Richard Titmuss was able to sum it up more frankly after the war was past.

'The same thing for each of some thirty cities,' Titmuss says, with 'the same monotonous and insufficient food in the rest centres, the same meagre provision of clothing, blankets and washing facilities, first aid, lavatories, furniture and information and salvage services'. He stresses also the inadequacy of unsupported assistance personnel and of too casually organized volunteers, as well as weak police liaison, and rounds off, a decade after the events: 'All these faults were constantly in evidence during the winter of 1940–41 as one city after another was bombed.' That puts it too mildly, is too kind in suggesting the lack of preparation was peculiar to the blitz winter – whereas it went on long after. Most impressive, in hindsight, is the persistent neglect in sectors where reorganization and general rethinking could have been executed without treading heavily on provincial toes. The incessant failure of the information services provides the clearest case. There was a new Ministry of Information, which, well conducted, could have intervened in local situations without annoying aldermen or challenging other extant vested interests.

What seem to be the most important factors which can help explain what frequently amounted to something like 'criminal negligence' – though by the nature of the British process, no one was prosecuted or purged, many of those responsible were praised, even honoured publicly?

The first reason why there was so pitifully little civic learning was undoubtedly because the citizenry of each place, including those who temporarily panicked, did by and large 'carry on', improvising individually. War production did not 'come to a standstill' after the bombing. Had it done so, Whitehall might have been led to intervene far more energetically.

Moreover, as a by-product of the same improvising ability, the miserable post-blitz services such as Rest Centres and food relief were rarely attacked publicly for their virtual collapse. The people did not rise up in wrath and lynch the mayor. This politically quiescent aspect of the public attitude, noted nearly everywhere – even under what practically amounted to provocation – derived mainly from the people's need to cope with their own problems, leaving little energy for anything else. Along with this, again, went the feeling that the whole shambles was an uncontrollable disaster, a distorted Act of God, involving everyone alike.

Second, everyone in the place was involved. Town Halls or Police Headquarters made larger targets than semi-detached villas. The experience was fully shared. And if a mayor here or there panicked, what street dared cast a stone? In the shared debris of the aftermath, also, there was no accepted standard by which the people could measure the emergency services, establish whether they were better or worse than elsewhere. Beside this was the regular post-blitz dramatization, the belief that no other place, no other people could have suffered worse than this, the most terrible raid possible. Suffering became commendable, if not actually heroic under these conditions of proud shock.

Third, outside London there had been a steady but extensive brain-drain from the smaller towns and cities into the few great ones; and from the great into the greatest, London – with intellectual offshoots in Oxford and Cambridge. Decisions and decision makers were overwhelmingly centralized. The war brought new ministries, which even drained Oxbridge: the great London University tower in Bloomsbury, for instance, pulsated with amateur propagandists as a Ministry of Information.

Fourth, provincialism was an essential structural part of Britain itself: in the forties far more so than in the reorganizing seventies. With roots embedded in feudal times, every unit of local government, large or small, was proud of its separate identity, defiant of any threat to infringe its integrity. Most of these units had evolved over centuries without relevance to the needs of the twentieth century. The resulting provincial patterns mattered deeply to local politicians and officials whose determination it was to perpetuate these relics from the past. From the existing, fragmented system these good people derived status, importance, outlets for service. They were concerned with their own local electorates, and were sometimes positively hostile to the interests of adjacent ones.

Fifth, the Regional Commissioners whose network was intended to overcome expected major breakdowns, had to be planned with full recognition of these restrictive local attitudes towards any form of outside intervention. The limits on their powers were in fact too strictly, narrowly drawn, although in Scotland organization of the District Commissioner was much more realistic. Insofar as the blitz was concerned, no Mayor or Town Clerk ever looked like giving over his local authority to the RC however great the mess. We have seen (in Southampton) General Hodsoll commenting on the RC's own feebleness when local officers had patently lost control.

Sixth, the Regional Commissioners, along with ARP in general, came under the Minister of Home Security at the Home Office. During the relevant period, this was Herbert Morrison, a most able administrator who made his reputation in local politics as leader of the London County Council, the largest and most efficient local body in Britain. He profoundly believed in local government.

Seventh, from Morrison down, politicians and administrators were most reluctant – if not actually unable – to admit defects in tried, existing systems plus their accretions designed for emergency. At any time, for such people openly to admit failure was pretty well impossible. In wartime, under new pressures, to make such an admission might smack of defeatism, just as to denounce the politicians might be treacherous. If this was self-deception, it was at least honest; and there was a lot of good sense in it too. It enabled responsible people to carry on with optimism under frightful pressure. But it could lead to disastrous complacency, as for instance in Liverpool. The same optimism was widely shown lower down the scale, with the majority in 1940–41 always believing they would not be blitzed, or that this raid would be the last, or that the war must be over in a matter of months. The only difference was that such optimism at the lower level was often despised from above as 'wishful thinking'.

Eighth, closely linked to seven, the image of a nation at war had to be one of unity, unfaltering courage, maximum effort, equal sacrifice and so on. Criticism from those not ready to take so much for granted looked 'unpatriotic', maybe even subversive.

Ninth, and behind all these broader factors, the same sort of evolution as had established the local authority's own 'civic sense' perpetuated another tradition: that the poor, the homeless, the deprived were to be treated as social failures, on no account to be encouraged to sponge on society. The blitz made misfortune and deprivation indiscriminate, unearned. But the organization and routine of public assistance was ill designed for dealing with situations requiring generosity, sympathy and spontaneity.

None of these nine factors automatically meant that *nothing* was learned and passed on. It was. Only outside London it was a very slow business, at a time when quick changes were essential. Notable here were the efforts made by several voluntary organizations to adapt more quickly. The Women's Volunteer Service, founded and run by the remarkable Lady Reading, was outstanding. In individual cases one noted major acts of initiative by the Salvation Army and the Christian Scientists, and sometimes the Y M C A.

It is also evident from the record that the technical side of the service improved faster than the psychological or social. In January 1941 fire-watching procedures were galvanized by official action; in the spring the provenly archaic system of locally autonomous fire services and the wartime Auxiliary Fire Service were at last merged into a National Fire Service. National shelter policy, forcibly modified by public take-overs in London, greatly speeded up the building of strong surface shelters and – as the year advanced – developed deep-boring plans, in a few places, like Portsmouth and London. The new indoor Morrison (table-type) shelters were moving into full distribution as the blitz faded. As Terence O'Brien's official history gently remarks: 'by the middle of May 1941 the government had made a substantial effort to fit shelter policy to the pattern of attack.' 'But,' he goes on, 'the more ambitious welfare activities had hardly begun.'

These 'welfare' activities were the subjects of incessant observer concern after the blitz. We made follow-up studies at Coventry (July 1941), Portsmouth and Southampton (through the summer), Plymouth (into winter), Hull and Liverpool. In June 1941, we wrote a general review 'Social Welfare in the Blitz towns' . . .

By July Coventry had provided 'reasonably adequate' public shelter facilities with amenities. Many of the earlier brick shelters in the streets had literally fallen down, owing to the usual defective lime-mortar. These were being rebuilt with 100% central government funding. Hutted camps for 10,000 outside the city were being planned as reluctant recognition of the reality of trekking, though this project was, according to local officials, being opposed 'from the direction of Whitehall'. Rest Centres were numerous and much improved, better equipped and backed up by a plan ('Come Right In') whereby registered households would take in bombed-out neighbours. (No organized help of this kind was met with in any actual blitz. The very idea of camping out with neighbours was seldom considered in the social structure of the streets of the forties.)

A first experiment in 'intermediate hostels' in Coventry, catering for difficult billeting cases, was just being launched in July 1941, to take large families, each one in a room with bunks and kitchenette. Three 'excellent' municipal cafeterias were serving good meals, with Sunday dinners a speciality – an ample helping of roast beef, new potatoes, greens, pudding, bread and marge and a cup of tea for one shilling. In conjunction with surrounding rural district councils 'a very sensible plan' was ready to tackle post-blitz 'refugees', with cooking depots set up in suitable villages, emergency accommodation on a regular basis in village halls. Much, in brief, painfully lacking before was now either prepared or being planned.

Although a Citizens Advice Bureau was working and seemed efficient enough, Coventry's broader approach to post-blitz information remained unclear. This report ended with words applicable all over for the summer of 1941: 'The view seems to be held in Coventry that there is bound to be chaos for a day or two after a blitz anyway, and nothing much apart from issuing posters can be done until people have recovered from their blitz-shocked condition.' There is quite a lot in this view. All depends however on the *degree* of chaos anticipated and accepted as permissible. Democracy with good government has to start from the assumption that chaos is not acceptable and is avoidable.

After the blitz passed, the slowest improvements were generally those made in the information services, Rest Centres, arrangements for emergency transport, trekker control (despite Coventry), communal feeding (often 'grossly inadequate'). The strongest link in the social service chain was emerging, early on, with the Assistance Board, though much depended on fresh, bright individual officials moving into difficult areas. Generally speaking – as noted in one of our reports of this time: 'The Board's attitude to the human need is that everyone who *may* need help is entitled to get it, at once, even if a few crooks benefit as well. The immediate cash advance to those bombed out, made by the Board, is probably the most effective, and indeed one of the very few existing contributions to a renewed belief in the future and the general goodness of society.' (FR 722). Plymouth's early achievement with Assistance has already been described.The city had made some other notable efforts by October 1941. Voluntary organizations and officials from all over Devon and Cornwall carried out a big exercise testing the capacity of the area around Plymouth for emergency dispersal, cooking facilities, transport to rural Rest Centres for the homeless and so on. We found that 'careful precautions had been taken to prevent a recurrence of the post-blitz confusion in official administration'. Parts of the machine which had proved least effective – information, billeting, Rest Centres – were being reformed. A bright young woman was appointed Welfare Liaison Officer, a new post to coordinate the services in place of their previous fragmentation. A Citizen's Advice Bureau was opened on 26 September 1941, including inter-unit representatives (billeting, war damage, assistance, education, the WVS), so that 'weary post-blitz trekking from department to department should not be seen again'. An 'Old People's committee' and two special 'Recuperation Hostels' (for those 'unbilletable through uncleanness' or nervous) were also in the works.[2]

Added up, these local efforts, though belated, became impressive. But they were left largely to local initiative. Had these and related measures been tried as need *first* arose, the successful ones then rapidly applied in all endangered communities, the strains of living through the blitz must have been enormously alleviated. Why was that not done?

Our nine restricting factors provide most of the local answers but not the full national one. The crude fact was that those at the top did not know what was going on at the grass roots outside London. Each local authority put a good face on everything. Each Whitehall department concerned at the local level did likewise. That was normal practice. What made the situation not so much abnormal as catastrophic was the absence of any analysis of how things were really working out, in human terms.

When public pressure to take over the London Tube stations was causing government concern, Mrs Churchill went out into the blitz one night and had a look for herself. She was appalled at conditions underground and hurried back to tell Winston something must be done for the shelterers. 'You work out something, Clemmie,' he replied. Lord Beaverbrook was recruited, made a second tour, confirmed the lady's shocked observation and pushed action through the Cabinet. Such was the random system of mass-analysis in 1940.[3]

The evaluation of what goes wrong and right in battle has long been regarded as essential in dealing with what Clausewitz called 'the friction of war'. By the same token LESSONS TO BE LEARNED is a key heading in *all* British military 'appreciations'. This approach was not carried over on to the Home Front, so wide was the dichotomy between civil and military outlooks as the war unrolled.

No routine had been planned to monitor the masses in spite of their alleged potential for 'friction'. This first became noticeable in the Ministry of Information, because this new body was primarily charged with morale and propaganda directed towards the home front. Soon it grew clear that this ministry was not fulfilling its intended role. As it was a new ministry it had no traditions or ingrained prejudices. It was, from the start, improvising, whereas the old-established bodies had already developed their war organization in times of peace. Although Home Security was also new, it was grafted on to its aged associate, the Home

Office, under the one minister. These older ministries were sure they knew what they were doing. They did not need anyone to tell them their jobs.

By the end of 1939, Dr Gallup's new offshoot, the British Institute of Public Opinion, and Mass-Observation (founded 1937) had been approached to check on the effect of the MOI propaganda campaigns. But the first minister, a colourless judge, vetoed both when he heard of the proposal. His successor, Lord Reith, already used to BBC Listener Research (started 1936), was more amenable. In April 1940 we received a regular contract to do 'both routine work and special crisis work' for the MOI, while retaining full independence. We thus worked for them all through the blitz.

In June 1940 a scientific effort to look at the air-war's probable coming impact was undertaken by Professor J. D. Bernal (a brilliant Communist crystallographer) and Dr F. Garwood (then attached to Home Security). Ironically, they chose Coventry as the model for a prediction of German bomb effects. A good first effort, it has received somewhat uncritical acclaim in the literature of operational research. Unfortunately these scientists – like everyone else around then – were preoccupied with major damage and serious casualties, having no word on the less obvious effects, temporary damage, homelessness, human disruption generally.[4]

Only when the blitz came could weaknesses of the statistical and physical approach be seen. The other factors had been consistently neglected. To try to fill the vacuum, the infant Home Intelligence section of the MOI, under Mary Adams, intensified its activities and widened its interests. This led to the Wartime Social Survey as a formal unit, precursor to the present Social Survey in Government. M-O's efforts were encouraged and elaborated by vigorous help from the Director of Naval Intelligence, Admiral John Godfrey, who became very concerned with the impact of the post-blitz on sailors' wives, ships' crews, shipyard and dock-workers. With the firm support of A. V. Alexander, the Cooperative Party leader and First Lord of the Admiralty, we were given *carte blanche* to operate independently and describe all we saw in these places.

This is not meant to suggest that MOI and DNI had developed good study methods by the time the blitz came. How could this be so when one basic theme, morale, was treated with such intellectual confusion? But, in the vacuum of neglect left by other and at least equally concerned departments, these two were at least trying. Nearly every record that has survived from those times came from these limited sources. Insofar as the data were inadequate, that is a measure of the disregard shown throughout the blitz, for discovering – let alone applying – LESSONS TO BE LEARNED from the frictions of total war on the Home Front.

But there was more to it than that ... other ministries and authorities directly engaged in post-blitz operations not only failed to analyse their own weaknesses; they positively objected to others doing it for them. No one likes that much, anyhow.

By the end of 1940 the Admiralty were near boiling point, seeing naval and dock personnel suffering great distress in the southern ports and on Merseyside. Alexander, with the help of the First Sea Lord and DNI, was pushing the blitz reports at other ministries and taking them up at cabinet level. The now opened files at the Public Record Office show some of the angry responses. For instance, 'Minhomsec' minutes to Parliamentary Secretary Mabane at the Ministry of Home Security, 21 January 1941, characterized our reports as 'a most extraordinary mixture of fact, fiction and dangerous mischief'. The writer 'presumes they were written by the intelligentsia' and advises 'the matter should be taken up *most strongly* with the Admiralty' – with a view, of course, to their suppression. Ten days later Morrison wrote personally to his socialist colleague and friend, Alexander:

> Mabane has shown me some reports on bombed towns which had been prepared by the 'Mass Observation Group'. I must say that I was a bit surprised to find that the Admiralty should have circulated such a document. To my mind it is very much like somebody with no naval experience or training circulating reports upon the morale of the Navy and its efficiency in action.
>
> I do feel that this document should be withdrawn from all those to whom it has been distributed and I hope *very* much that you will arrange for this to be done.

Alexander replied with naval gallantry, refusing to withdraw. On the contrary, this activity was stepped up. Minhomsec was provoked to further protest to his Minister (27 January); and so on – it is another story. Home Security and other less important ministries treated it all as bolshy criticism, tinged with subversion. Ministers held firmly to the traditional politico view expressed by Morrison in his autobiography, where he claims that 'one thing I did from the outset, and that was to go out and meet the people, to see things on the ground and talk to humble civil-defence workers'. From the record it looks as if Lady Churchill and Lady Astor did a better job of seeing behind this pastoral humility. No observer found any trace of ministers other than Churchill (and he briefly) really getting on to the ground, away from their own officials, and advisers, to look and listen. In any event, a great machine needs to measure its own effectiveness far more carefully and impartially than that ... as, nowadays, all would admit, no doubt?

As the blitz spread wider across the land in 1941, other ministries did begin to feel the need for some better internal system for learning by experience. Only in March 1940 had the Whitehall end of the Ministry of Health, with its multiple post-blitz responsibilities, started to receive systematic reports from its regional officers; (to quote O'Brien) 'long after this date this particular problem persisted.' Exchange of intelligence from periphery to centre was minimal. Because London absorbed so much blitz attention during late 1940, 'a process of critical *local* surveys by Ministry of Health inspectors did not get under way until the beginning of 1941' (Titmuss). Although the reporters were themselves career civil servants, their reports immediately revealed 'many glaring inadequacies'. Yet even after these inadequacies had been pointed out, 'it was rare for any city to take emergency action'.

Similarly, the first circular issued by Health to provincial authorities on the need for information about administrative centres was sent out on 16 January 1941 – the reader may remember that the Home Intelligence reports had been hammering away at this point from the start. And regular, standard reports from the provinces to Whitehall on this and other matters of moment were not established as routine until July 1941.

Outside London, there was similarly no regular inspectorate of air-raid shelters, though for civil defence on the ground generally the dedication of Wing-Commander Hodsoll, as Inspector General of the ARP, was crucially important – in as much as single voices inside ministries could influence the whole interlocking picture.

In Scotland, usually ahead in this sort of thing, active learning began after the Clyde raids of March 1941. In the following months the Scottish Department of Health in particular developed flying squads to assist locally, while inspectorate services were much improved, inter-authority cooperation evolved, and so on.

In July 1941, a special report, 'Preparations for Heavy Air Attacks next Winter', was prepared inter-departmentally for the War Cabinet. This – wrongly – anticipated renewed blitzing in September. Many earlier breakdowns were now clearly pinpointed and set out in a high-level decument, with special emphasis on the need for inter-authority cooperation and joint planning committees from all departments – and on the importance of information to correlate the lot.

Important in another way was the August 1941 paper based on post-blitz studies in Hull and Birmingham which was circulated by Home Security and marked another step towards 'humanizing' the approach. This study examined the vulnerability of humans to bombing and concluded that one needed larger quantities of HE to affect industrial urban activity seriously. But the report paid little more than lip service to the full complexity of morale factors. Next month, the research officers responsible in Home Security planned a 'correlated analysis' study of raid effects on production, transport and the 'general morale of workers' in Hull (still being bombed) and Birmingham (as an industrial centre). In October the Air Ministry lent its support to research which, it believed, would assist in directing their offensive against the heart of Germany: 'It is hoped that this study will enable an opinion to be formed as to the possibility of bringing about a decisive breakdown in morale in urban districts as the result of air attack.' (AM to MHS, 11 October 1941.) The work proceeded in Hull and Birmingham and was circulated in a summary version in May 1942 by the Research and Experiments Branch of Home Security. Despite an explicitly narrow interview approach, this paper was titled 'The total effects of Air Raids'. An earlier fuller version apparently went to the Air Ministry whence Portal wrote enthusiastically on 23 April:

'The report will prove a great help to the Air Staff in planning operations.' The broad inference was simple, namely, that it might after all be possible to shatter civilian and specifically worker morale by bombing, only one must do *much more of it* than the Germans had. This suited the approach favoured by 'Bomber' Harris, and so near to the Boche-hating hearts of Cherwell and Churchill. Yet the study was in some respects incomplete, intellectually just as imprecise as the rest of us at the time, and in the outcome potentially erroneous therefore. For here we come to the final puzzle of non-learning: that having demonstrated what and how humans could survive and adapt to in 1940–41, some of the best brains in Britain re-examining the evidence felt able honestly to indicate that it was all a matter of *quantity*. The Germans, under heavier attack, would crack. Although they did not, the same premise has since been at the base of the vast aerial activity of the Americans in Indo-China. Other humans – of implicitly inferior race – *must* be subject to Douhet's law of civilian vulnerability. Indeed, to believe otherwise might well be to deny the very significance of being an airman.

To go further on this theme is outside the bounds of this book. But one paragraph from the Hull–Birmingham report may serve to illustrate the war-conditioned undertones: 'To attack the population, houses must be destroyed, and for this it is most economical to attack the densest zone, which is invariably in or around the city centre, that part of the town which is most vulnerable to fire. Thus, in the attack on people, our present technique of firing the city centre to provide a beacon on which later arrivals carrying H E can concentrate is perfectly correct ...' The 'argument' here is open to flat contradiction. The very nature of civilian war may be misunderstood in this, the accepted, aggressive attitude.

Be that as it may, although the scientific assessment of human reactions was making some strides as the war crashed on, data of sufficient quality and quantity were not available in most cases; thus the lessons to be learned from such assessments were astonishingly slow to be implemented. Most of the lessons were learnt (if at all) only after 'our' blitz was over. Democratic leadership in Britain failed to live up to its domestic obligations. The peacetime social and political structures were too rigid, too self-satisfied and insensitive, too sluggish.

11 Human Effects

> The redeeming feature of war is that it puts a nation to the test. As exposure to an atmosphere reduces all mummies to instant dissolution, so war passes extreme judgement on social systems that have outlived their vitality.
>
> KARL MARX, 1897

Many immediate, short-term, visible effects of the blitz on Britons have been described in the preceding chapters, from popular panic to personal pleasure. This chapter aims to sum up these as well as certain less transitory effects, some of them difficult to see, others seemingly significant at the time but now, when seen in perspective, of little real importance. We may also venture to generalize a little on the overall influences, if any, of blitzes on those living through them. When war passed judgement on our social system, how did its vitality measure up in terms less obvious than the temporary breakdown of the administration machine? Two particular effects are considered first because these were very generally thought to be the two which would demonstrate most strikingly the demoralizing impact of bombardment.

(1) Industrial effort

The original Douhetist idea of bombing aimed to shatter civilian life is not necessarily 'wrong' because it has generally been proved so by the costly but incomplete experiments so far conducted. During the Second World War it was for a while modified, from British experience, into the belief that indiscriminate scattering of bombs on human concentrations would at least disrupt work patterns. But, as we have seen for Coventry, Portsmouth, the Clyde and other places, the evidence was in part misinterpreted, in part inadequately re-examined during limited post-blitz studies.

After that war was over, Richard Titmuss, looking at the same information as enabled British scientists at war to encourage Bomber Harris in his dreams of disintegrating Germany, pointed out that even a worker whose home was rendered permanently uninhabitable only lost on average six days from work. The officially deplored 'trekkers' lost no more work-time than those who continued to sleep at home in target areas, as was equally Mass-Observation's own experience.[1]

The *need* to work for money, linked to the habit of working (and the fear of not having a job) proved far more powerful than the prophets ever supposed, not only in 'essential industries' but – as some of the diarists have illustrated with beautiful simplicity – in the cases of shopgirls, secretaries, milkmen ... Birmingham was a prime target for Luftwaffe attacks on industrial output, but heavy raiding there achieved only a temporary reduction in production of about 5%, while firms not actually hit sometimes showed no decline whatever. BSA in Birmingham were actually able to increase their production of Browning guns from 894 in December 1940 through the heavy raids to 3750 in March 1941, though their rifle production at the Smallheath works did fall sharply for a time.

The two main analyses of manpower use and labour in the munitions industries both treat air-raids as secondary causes of work-loss when compared with a multiplicity of other factors. One does not even mention bombing or the blitz in an exhaustive index.[2]

Much was made at the time of the disruption of German industry. The exhaustive four-volume study of air bombardment by Sir Charles Webster and Dr Noble Frankland (now Director of the Imperial War Museum) reviewed all the allied and internal data which became available after the war. Bombing of Germany was, after 1941, enormously heavier than anything the Luftwaffe did to Britain. Berlin received more than twice as much tonnage as London from the RAF alone and Frankfurt nearly twice as much. Augsburg in a single night got more than Plymouth and Bristol in the whole war. Nearly half a million tons were dropped by Bomber Command on to Germany, 99% of it after May 1941 – as compared with well under 40,000 tons during the blitz on Britain. If we take munitions output in Germany with a January–February 1942 index figure of 100 – before the great assault unrolled – for the whole of that year the comparable figure was 142; in 1943, 222; and in 1944, 277. Weapons output had trebled by then and the production of panzers multiplied by more than 5. While in 1940–41 together Germany produced 21,000 military aircraft to Britain's 25,000, in 1944 Germany completed 40,000 to Britain's 27,000. These comparisons are crude; but they do serve to indicate the trend.

The estimated total loss attributable to all urban attacks on Germany, allowing for the lag in effects on industry, was finally calculated at 3·7% of total war productive capacity (excluding mining), and fell to 1·2% if metal and iron processing were excluded as not being wholly war production. Estimates for various German cities showed Hamburg, with the highest 'index of density' in bombing, as losing 3·4% of the city's contribution to overall German production as a result of this huge blitz. Raids on other industrial centres like Dusseldorf and Bremen gave much lower figures. From all these data, largely the result of the U S Strategic Bombing Surveys, a rough estimate indicates that for every 15,000 tons of bombs there was a loss of about 1% of annual production. This at a cost of over 8000 R A F bombers and well over 50,000 young, highly trained personnel killed – more than all those who died in the whole blitz on Britain.[3]

None the less, in October 1943 'Bomber' Harris was writing at length to his Minister generalizing the claim that '8 months intensive bombing' of Germany could now 'lead to her collapse at a very early date'. He went on to complain bitterly that the government was not being open enough in declaring itself on this matter. It 'should be unambiguously and publicly stated', he urged, that 'the destruction of German cities, the killing of German workers and the disruption of civilized community life throughout Germany' formed a key part of Britain's strategy.[4]

(2) Psychiatric and shock impacts

As we know, the predicted crack in nerves did not send a large slice of Londoners gibbering in September 1940. The hospital beds and poised psychiatrists waited, unemployed.

The record shows plenty of personal, temporary panic as well as flight. Nowhere can one identify – either from subjective personal reports (including those from some very highly-strung people) or by objective description – an individual who because of blitzing, became any kind of lasting hysteric, or any group which became 'uncontrollable' in shock for more than a matter of minutes. Fear and bewildered anxiety were greater then in the wartime propaganda stories; and these heroic stereotypes have confused the journalist and historian. But from the Mass-Observation side we can report only the mildest defiance of any policeman and nowhere a collapse of human relations, or general lawlessness. It is important to restate this part of the blitz story, especially because the subsequent picture has been distorted one way, while the earlier picture – predicting the future – was twisted in the opposite direction.

Reinforced perhaps by those excited reactions to the first siren, the most fashionable of London psycho-analysts was able to write as late as 1940 of 'the utter helplessness of the urbanized civilian today', so that 'impotent fretfulness' must lead to wide-scale collapse under the strains of total war. Shocked citizenry would seek psycho-escape into 'infantile security'. This became trade dogma for anticipated blitz reactions: people reverting to the pram, the womb or the tomb. The view gained further support when unexpectedly many of the September 1939 evacuees from town to country were found to be chronic bed-wetters: enuresis, reassuringly close to full neurosis.

There seems to be only one published general study which clearly might have given other clues. Two Americans, E. Whitt-kower and J. P. Spillane, writing in 1940 from a re-examination of the scanty records for the First World War, qualified the familiar reports of panic in London's East End. They found these superficial, in that London hospital patients there had shown 'an absence of any marked (nervous) reaction'. They went on to note, more widely, that 'the civilian population stood up surprisingly well to the terrifying experiences of repeated air attacks', especially during the Spanish Civil War.[5]

So, the leading British Freudian (and member of Home Intelligence's Morale Committee), was to find 'monotony' once the reports on raid effects started coming in: the monotony of nil reports. Dr Glover was able to tell a meeting of fifteen practising analysts later in the blitz that only one 'genuine' breakdown case had come to his notice. Less than three nerve patients per week appeared on the couch over the first three months of London's bombing.

Dr (now Professor) P. E. Vernon recorded under 1·5% out of more than a thousand London shelter hospital patients with any sign of nervous trouble. At a specialist London hospital up to 2·5% of the intake suffered from a malady which could be in some way related psychologically to air-raids. Overall, however, nervous admissions dropped with the end of peaceful nights and then went down further in 1941 – as did female suicides during the blitz years.[6]

The outstanding study on this theme was carried through after the war by Dr Irving Janis of Harvard, partly in a survey sponsored by the Rand Corporation which flatly decided: '... neither neurological diseases nor psychiatric disorders can be attributed to nor are they conditioned by air attacks.'[7] There were, naturally, many less clear, less emphatic phases of mental disturbance, including anxiety, temporary speechlessness, more protracted apathy (among older people and women), acute pessimism (though more often the opposite), some wildness in the young. Little of this was peculiar to the blitz alone. On the whole, the blitz in Britain tended to reduce fantasy anxiety states. In 1975 there are still people who shudder every time they hear certain sounds which recall sirens, and so on. But these are smallish things, not in the same class as trouble with your bank manager, mother-in-law, or wisdom teeth.

(3) Religious

To many people the blitz looked so cruelly indiscriminate, so 'senseless' as to call into question the purpose behind anything in higher western ('civilized') ethics. This was a true and serious 'shock' brought about by bombing experience.

It is difficult to put a finger on these spiritual factors, more evasive even than 'morale'. But often at the time and sometimes looking back again this writer has the distinct impression that the heartlessness implicit in the new methods of war made ordinary people re-assess standards of public morality previously taken for granted in this island – where a single spinster executed (after trial) by the Germans in Brussels in the First World War rates heroine status with a statue in the heart of the West End. It may well be that the blitz did a good deal to accelerate that devaluation of the orthodox Christian code. It also assisted the growth of alternative humane philosophies.

The harsh spiritual implications of the blitz, not only on us but *by* us, might well have been softened had the churches played a larger part in providing physical and intellectual leadership. They seldom did so. To an extent that remains surprising, churchmen regularly failed to respond, in public anyway, beyond the necessities of their own narrow groupings; frequently they failed these too. There were exceptions. But the fact is that, in writing this, one almost instinctively wants to insert the word 'notable'. The exceptions were so notable because so few. Though no greater than the simultaneous failures of the local politicians, it could be said that for churchmen, with so few administrative responsibilities, the blitz brought new, great opportunities, a major challenge. The challenge was not taken up. Even undamaged church buildings regularly stayed empty (if not actually locked). Nowhere is there any sign that many turned to prayer in these civil crises.

A 50-page M-O report of youth in London's blitzed winter ('Young People', January 1941) gave 1% after-school teenagers participating in some form of church or other religious activity at weekends, none during the week – compared with 34% and 15% respectively for cinema-going, and nearly one in ten doing some sort of leisure education (night school and the like). The figures are at least indicative of a trend which continued beyond the war itself. The overall situation was fairly put in another report of January 1941 ('God and the People') where the writer ended on a more personal note regarding the church at that stage in the blitz:

The façade stands. Behind it, the roof is crumbling into the chancel. I have had occasion to visit every one of England's blitzed towns in the past month. Sometimes I have wanted to go and sit somewhere quietly and think. In Portsmouth at the weekend I tried eleven different Churches and Chapels. Not one was open. Then at last I found one where the knob turned to my hand and the door opened. I stepped inside – and nearly stepped out of this life. Destruction had been there before me, and the façade laughed in my back.[8]

There were some minor signs of a religious revival after the blitz, as the war went better in 1943; some strengthening of faith among those *already* deeply religious, especially women. Alongside this, among the non-religious appeared 'a desire for new standards and values of a religious character' without 'in any way turning to religion'. The subsequent frustration of this desire was another casualty of the Second World War.

(4) Social change[9]

In terms of psychiatry or spiritual belief, there is no marked distinction between short- and long-term effects derived from living through the blitz. In terms of social life, habits and organizations, this distinction is more apparent, with immediate effects numerous but not necessarily lasting; and largely mixed up with other effects of the war in general, not peculiar to the blitz.

The best information available to put this aspect in perspective comes from men and women in many different situations, scattered across the United Kingdom as Mass-Observation's observer panel. These were not a 'statistical sample' of the population in the modern psephological sense, but they were candid, keeping monthly records which give information no longer available anywhere else. Soon after the blitz ended, in June 1941, these people were given a list of 27 activities, selected from previous checks on habit and mood changes. They were asked to indicate whether they were doing less, more or the same amount of each. Men noted on average 14·9 changes each, women 15·2. The most interesting result here is the difference between blitzed towns and other places, a difference significant because it is so small, only one month after May:

Average number of changes per person	
Blitzed	15·1
Other towns	14·7
Rural areas	14·4

With or without blitz, the enormous amount of change emerges at once. Of the 27 items listed, *none* were mentioned by less than 25% among the men, and only two ('drinking' and love life – with related anxieties) among the women. Changes in travel habits, home patterns, contacts with close friends involved three-quarters of both sexes.

Secondly, the change was mainly negative: to do *less* of what one did before, not more. Very little was done that was new, except insofar as a household habit would be carried over, modified, into the Anderson shelter, for example. People did not tend to take up cards, skipping or whisky if they had not favoured these outlets before. Rather, they were inclined to do some of their old things more, more of them less. *Much* less done were travelling beyond necessity, seeing friends, cinema going (53% went less often), other outside recreations (48%) and sleeping (47% reckoned they had less sleep than before). Dancing scored 27% less and 7% more – we have seen it as among the most persistent social activities in blitzed towns, such as Southampton and Liverpool.

Smoking and drinking both, exceptionally, were on the increase especially among males; 20% more to 12% less drinking, 33:11 for smoking at a time when supplies were difficult. Three things were done – very significantly more by blitzed people than others.

	Ratio of men doing this more in		
Habit	*Blitz towns*		*Countryside*
Smoking	5·7	:	1
Sleeping (or trying to)	2·4	:	1
Reading (books mostly)	1·6	:	1

The other way round, three things were significantly more in the country.

	Ratio of men doing this more in		
	Countryside		*Blitz towns*
Seeing friends	2·1	:	1
General recreation (sport etc.)	1·6	:	1
Cinema going	1·5	:	1

These trends could to some extent be measured against a 1937–8 survey from the same observer panel on the same subject. Smoking showed up most clearly in this, The majority of pre-war 'chronic' smokers were smoking through and after the blitz. During blitz periods a marked increase was noticed among A R P personnel with long periods of more or less anxious inactivity. Pipe-smoking went up most in these situations:

Man: 'On A R P duty I smoke more, chiefly to keep awake.'
Woman: 'In London during the heavy raids I found smoking a great help.'
Woman: 'My consumption has gone up over 100%. It started in September during the blitz. I found smoking kept me from getting jittery.'

Differences between the sexes were summed up at the start of summer 1941: 'There is some indication that women are adjusting themselves to more radical changes at least as quickly as men. Fewer of them are externalizing the tensions they mention in annoyance, they are apparently more often completely occupied than men (i.e. don't find themselves at a loose end). Nearly a fifth of them feel happy more often than before the war, compared with a seventh of men.' It was difficult to sort out exactly what was a change due to the blitz and what not, even in the big cities. Clearly, also, as the blitz died away in May, some of the 'new' adaptations faded back into old tricks.

A wider idea of pre-to-post blitz trends can be gained by comparing the observer panel over all the war years. Some observers fell out, new ones enrolled, but by and large there was continuity; and again, there is no better source (alas). Each December the panel reported the inconveniences they felt most at that time. At Christmas from 1939–40 to '43–4, the blackout and transport troubles filled first and second places. Most people never really got used to either. Blackout came first every time but one. That single fact quietly but forcibly demonstrates the extent to which people living through the whole war longed to be rid of most habits imposed thereby. Blackout began the evening of Neville Chamberlain's radio announcement and never let up. Ostensibly, everyone in Britain adjusted to it. Yet they never got to take it for granted. Adjustment might mean acceptance; but it did not build up fundamentally new approaches to life. Taking the top six for the first pre-raid winter, just to glimpse these trends:

Inconvenience	*Order of importance in:*				
	1939–40	*1940–41*	*1941–2*	*1942–3*	*1943–4*
Blackout	1	1	1	2	1
Transport	2	3	3	1	2
Prices	3	4	8	10	8
Fuel, petrol	4	–	9	6	8
Food	5	2	2	3	4
Evacuation	6	13	11	19	21

Seen this way, the new habits learned and old ones changed through the blitz fall better into focus, as one part in a long process of civilian war.

The war changed some ideas and habits, in the long-term, though probably not so many or so much as had been supposed. The blitz on its own changed few. Much has been made, for instance, of the drawing together of strangers. This certainly happened in the big shelters like Tilbury. But it would be a mistake to make too much of these temporary associations. There were few signs of any keen urge to share once an immediate threat was past. There were as many fresh disputes and frictions as new fellowships.

Socially, those changes made necessary by living through the blitz were nearly all transitory, to meet inconveniences, tensions, survival needs. Once the pressure passed, the survival value vanished, they were easily discarded. There is no evidence to suggest that Plymouth trekkers fell in love with Dartmoor and decided to take to country life in peace, any more that Plymouth's people formally remembered 'Stainless Stephen', their blessed music-hall comedian, in the way that those who survived the

Burma Campaign rendered homage to this comedian in ten post-war Burma Star Rallies at the Albert Hall. There is no record – and no probability – of the veterans of Tilbury or Chislehurst Caves convening annually to celebrate what was, in fact, an enormous if amorphous social achievement.[10]

Entirely different was the material effect of the blitz, affecting the structure of British cities, their civic and shopping centres, their slums and crowded streets. Of some 13,000,000 private homes in 1939, nearly 4,000,000 were damaged (1,500,000 in London), about 220,000 totally destroyed. Massive rebuilding resulted after the war. But again it must be remembered that this was nothing new, only the continuation of a trend which caused 245,272 houses to be demolished or closed from April 1934 to March 1939, over a million persons moved without a bomb to speed the process.

(5) Political trends

Continuing a little longer with the panel of observers, they were also used to identify political and related changes which people expected to emerge as a result of the new, 'total' war. In the middle of the blitz they volunteered the six changes they considered most important for the world they hoped (or feared) would come out of present distress.

	*Percentage expecting this as major post-war trend** *(in January 1941)*
Less class distinction	29*
More state control	21
Education reforms	19
Levelling of incomes	15*
Increased social services	*14
Dictatorship, 'fascism'	13

Of these six, all but the last came to pass! Other changes less often mentioned included religious revival (9% of women), revo-

*An asterisk before the number means especially mentioned by men, after the numbers by women.

lution (5% of men) and better rural–urban understanding (8% of each sex). Only 6%, mostly women, foresaw higher taxation; none inflation.

The blitz played a part in realizing some of those expectations, mainly as one impetus towards equal sharing of war's burdens. But the big drive for major change came later, influenced largely by the length of the war and the great numbers of people eventually involved in it. The war had to be made worth-while to offset boredom not bombs. This mood was first crystallized in the Beveridge Report, nearly two years after the blitz.[11]

While an intangible increase in general self-confidence was one effect of the blitz, growing out of previous often exaggerated anxiety, it cannot be over-emphasized that nearly everyone tended to attribute this feeling to their own personal, practically unaided efforts. There was little to justify increased confidence in the political system, national or local. Yet there was exceedingly little inclination to 'blame' it, none to regard it as having 'outlived its vitality' (in Marx's phrase) in any long-term sense. The ordinary citizen, even where he might detect the structural faults, did not normally deduce any permanent inadequacy. In a way, this was another form of non-learning.

The blitz did not, therefore, arouse any wide or lasting antagonism to the leadership, although occasionally it accentuated existing local cynicism. Disillusion about the working of democracy was already well-established before the war; and what has been happening since in Britain and other countries is part of a long process of malfunction which was also demonstrated by the aforesaid civic breakdowns in the blitz.

Up to May 1941, 15 out of 22 by-election opponents of the political *status quo* were openly anti-war. By 1944 Common Wealth had 321 constituency branches and three elected M Ps. But only one of their candidates survived the revival of old *established* parties at the General Election of 1945, which threw Churchill out and Attlee (Labour) in.[12] Churchill himself received vast public and published acclaim as Britain's successful wartime leader. But the basic British distrust of strong leadership showed even when he appeared, crying, at the scene of blitz destruction. Millions in fact thought of Churchill, specifically, as a mighty support in dire necessity, a sort of intellectual deep-shelter, intended for emergency protection only. Once the war was won . . .[13]

Living through the blitz had no more measurable long-term effect on political thinking than on social habits. It just was not that *sort* of experience. It was too personal, too far out of the normal world of everyday politics, clubs, societies and society.

(6) International and 'racial'

There is little doubt that Churchill, his closest personal adviser Cherwell and 'Bomber' Harris relished the opportunity to destroy German cities and civilians. Their actions were sanctioned by Churchill's closest publicity man, Lord Beaverbrook, who vigorously reported and elaborated on public demand, nay clamour, for 'reprisals' through his news chain. The 'People of Coventry' and all the rest were thus said to cry out 'bomb back . . . kill . . . destroy' in vengeance.[14] Under such alleged insistence the leaders could only, in decency, bow to the popular will. The BBC caught it perfectly in the opening paragraph of the 6 p.m. news bulletin on 16 November 1940: 'This evening's best news of the war in the air – particularly for the people of Coventry – is that squadrons of British bombers last night made a terrific raid on the city of Hamburg. It was very satisfactory to the Royal Air Force men who carried it out.' These clarion calls failed to produce anything approaching unanimous demands for reprisals, even in *public* opinion (what you'll say to a stranger). Certainly all sorts of people did react to the blitz by calling for retaliation; and to say so became 'the done thing' through all the publicity media. Champions of reprisal were conspicuously more numerous in the unblitzed places and in the countryside, as well as among elderly males who had fought in the First World War. But one was repeatedly impressed by the paucity, sometimes the total absence, of such reactions among most blitz victims. The published versions greatly exaggerated the private mood.

The simplest single attitude may best be expressed as a combination of three reactions: (i) the misery, mixed with excitement and sense of achievement in being bombed – so what earthly use inflicting the same on non-belligerents on the other side?

(ii) the evident futility of so much lethal effort – why waste our own planes and lads to the same clouded purpose? (iii) the strongly felt probability that the more 'we' bombed 'them' the more they would bomb or in some way retaliate on us (even unto gas) – so why make it worse for everybody, ourselves included?

Many exceptions apart, the low intensity of demands for reprisal among those supposed to be justifiably clamouring for revenge stands as another indication of the extent to which the civilian war was felt personally rather than socially, politically or even 'nationally'. There was no massive ground-swell of opinion against anyone.

At Merseyside and Manchester we earlier noticed that 'no mention of reprisals was anywhere heard' during the whole period of blitz observations in late 1940 and 1941. So marked a negative was perhaps exceptional. There were also changes of opinion from month to month, even raid to raid, as also person to person. Thus a Lancashire man who had been shouting for reprisals in his diary, is later found reverting to his usual mildness:

3 a.m. warning. I hear that during the 3 alerts Nazis were over the steel mill district mentioned. They are trying to find the huge munitions works (hidden underground). When they do they'll send a fleet of bombers. But I don't know why our nights should be so hideous. Has it occurred to those who call Britons courageous and brave for carrying on in the blitz that Britons have no option? We must work to live. If the Government is unable to stop night attacks we must be content and be murdered. Reprisals are stupid; the only way is to stop the Nazis bombing. It can be done – we invented the tank. (20 November 1940.)

The term 'reprisals' itself became somewhat fogged in popular usage. Many took it to mean any sort of retaliation, including those against German armaments or Hitler's mountain lair at Berchtesgaden. A lot of those who verbally favoured aggressive reaction were confused. In Portsmouth, half those favouring any kind of aerial counter-measures said it would make them feel better if Germany suffered correspondingly. Three quarters of the population held no such views in March 1941, when 'reprisals featured as an insignificant part of Portsmouth's intellectual landscape'.

In April 1941 the British Institute of Public Opinion took a national sample on this issue. Like us previously, they found that the main demand for retaliation came from the least bombed parts of Britain. 'It would seem' (they said) 'that sentiment in favour of reprisals is almost in inverse ratio to the amount of bombing experienced.' So it was. In regular London street samples M-O mapped a big increase in reprisal demands once the blitz was passed. It reached a high of 67% by September 1941. Even then, there was much confusion: 'I *certainly* think we should bomb military targets.'[15]

Hatred there was – some. Surprisingly little of it was engendered by personal bombing experience, however. Mixed with this came, persistently, a form of affectionate respect for the Luftwaffe crews, 'only doing their jobs', etc. For every one reference to a Hun or Boche – Churchill's standard epithets – there must have been getting on for ten thousand to Jerry. 'Jerry', as Eric Partridge's *Dictionary of Slang* aptly remarks, 'is often half-affectionate.' It was not uncommon for blitzed people, especially women, to express a measure of sympathy for those risking their lives to bomb them. In the same spirit the Führer was commonly referred to (in private) as 'Old 'Itler' or 'Adolf'.

A traitor is supposed to be the most horrible product of war. Yet all through this one, the words of William Joyce were heard with attention by millions of loyal British citizens. Lord Haw Haw was Hitler's verbal projection into Britain. When the dissemination of information broke down during blitzes, his authority was considerable if confused. His importance had been obscured because officialdom felt he could be ignored or laughed off. Lord Beaverbrook's *Daily Express* launched a massive attempt to kill him by ridicule. Smith's Electric Clocks used him in their advertising (1940). George Black put on a (successful) musical at the Holborn Empire, 'Haw Haw'. A British Institute of Public Opinion Poll confirmed Mass-Observation's own figures for mass listening. People listened with a fine mixture of trust and doubt, laced with a sort of sneaking affection:

1. 'He sounds such a nice man. It makes you think there must be something in what he says.'
2. 'Although he appears to be an unpleasant personality I feel that there is something vaguely likeable about him, but this may just be due to all the fun that is made of him over here.'
3. 'I love him and his clever, tricky sayings. I love his voice and manner

and would love to meet him. I feel he is a gentleman and would be a nice friend – all my intuition and his voice . . .'

For a while, in the 1940 lull, interest in Haw Haw fell, though well over a million Britons stayed with him every day. In the blitz he became a voice of authority, though not perhaps nearly so much so as he could have been, more cleverly deployed. And if only a few people in Plymouth relayed his subversion (in good faith) the word went all round. On the side, too, he kept up the anti-semitism always latent in some Britons, although this played no significant part in the blitz.

This love-hate ambivalence (the affectionate enemy) comes bursting out in the hinterland of dreams, where people's more private opinions can rebound upon themselves. Some of our observers kept dream diaries, some were asked to report their dreams during special periods – with rich results which can only be briefly glimpsed.

Raid dreams were shared by old and young, but experienced more vividly by women. A young actress in London dreamed one raid: a German pilot came flaming down, 'a horrible mess of flesh and bones'. Then she was caught by a parachutist. When he removed his mask he was 'very young, about 17'. She smiled at him, stroked his head.

A spinster librarian in Devon previewed the dramatic Scottish landing of Hess by dreaming that she promised to shelter a German spy. He duly landed beside her house. 'I tried to pretend I did not know anything about his coming. But he came and beat upon the door.'

A young civil servant in the Midlands had a dream argument with Hitler, ending with a long harangue on the rights and wrongs of total war. Hitler (he found) 'was fairly humble but obviously unconvinced'. A Dorset housewife went walking down the garden path with Hitler, who promised her some nice plants. That same night her best friend dreamed she was in bed with Hitler, 'who had his boots on'. A Nottingham upholsterer met Hitler among a crowd of diplomats, went forward to address him, was disappointed he didn't resemble his photographs. Much later in the war an Ipswich man dreamed that Hitler came to tea: 'I wonder if he would understand what I meant if I asked for his autograph.' In August 1942 a gentleman living near Tring in Hertfordshire twice dreamed he was with Hitler; the first time 'he was quiet and almost subdued', the second he asked the Englishman for some books.

Mutual respect was usual in these nocturnal relationships, as again with an Essex widow (69) who was anxiously 'trying to tidy up a room . . . because I felt Hitler was possibly coming'. A retired schoolmistress in London found herself in Germany – as many war dreamers did – so close to the Führer that she was suddenly kneeling 'and looking up into his face with a feeling of devotion'. Then Adolf leant down and whispered to her, sibilant: '*Arthur is in this.*' At this she felt 'dreadfully disgusted'. No wonder, Arthur was her trusted brother!

A young woman on the south coast one night visited Brighton to find a dream-sky filled with German planes and airships 'like flies', led by a huge flying raft with six propellers upon which stood ('legs astride and hands on hips') the German High Command. One man 'well to the front looked like a Prince in a fairy tale'. Soon the ground was alive with Jerries, her mother announcing 'we always leave our front door open at night'. And to top them all was this retired London grocer: 'I dreamt that Hitler was going to be hung, and that I was in charge of the proceedings. There was a carriage and a troop of soldiers waiting for me – but I was worried as to whether I should wear a lounge suit or evening dress. I asked someone who said "Evening dress certainly".'[16]

Summing up, the blitz confirmed hatreds already felt by some, extended others, leaving many more barely touched. Maybe half of this third category at one time or another favoured hitting back, more or less tit for tat; but rarely was this view held with any enduring ferocity. It might get as far as tooth for tooth, seldom eye for eye, especially among those who had known the heartburn of actually being blitzed, and so long as this experience was still fresh.

Among the Allies themselves, there were not a few times and places inside the United Kingdom where, later in the Second World War, Britons hated Americans considerably more than they ever came to hate Germans, let alone Italians or Japanese.

The blitz did little and the whole war less to make English,

Welsh, Scottish and Ulsterman either hate their continental enemies or love their foreign friends. In Brussels in the First World War, the British hospital nurse Edith Cavell was shot at dawn for abetting the British cause in Belgium. It could be that this one act of 12 October 1915 did more to make the British dislike the Germans than the killing of more than 30,000 times as many British women and children in 1940–41. It may be seen as some kind of minor triumph for human decency and adaptability that by 1945 millions of Britons – if not their leaders – acted as if they could easily live with the nurse's last words which in the thirties rang a little false on the cement plinth of her statue outside St Martin's-in-the-Fields:

'PATRIOTISM IS NOT ENOUGH. I MUST HAVE NO HATRED OR BITTERNESS FOR ANYONE.'

(7) The young and post-war generation

Children have not featured largely in the foregoing descriptions. Most were either evacuated before raiding or left quickly once it began. Moreover, this work was necessarily most concerned with reporting upon systems and developments where adults had all the decisive roles.

Evacuation, country life and general war dislocation certainly played a larger part than any blitz, where the young were concerned. Of those who stayed put with their parents, a few were continually nervous and a few constantly exhilarated. The greater part adjusted as well as their parents or mildly 'better'. At no stage did they present a special problem as compared, for example, with old ladies or stray pets.

This very broad conclusion was confirmed by the psychologist P. E. Vernon in a post-blitz review of data from all sources where he concluded: 'All observers seem to agree that raids have even less effect upon children than adults.' The chief psychiatrist of Guy's Hospital in South London could not find a single case of 'acute emotional reactions among children' bombed. A Surrey Child Guidance Clinic recorded two cases of 'siren-fright' where the children had not been subject to bombing.[17]

In the nineteen seventies it is easy enough to blame the blitz or the war as a whole for disturbing those then young – and thus to help account for modern violence (as for instance in a *Times* correspondence of August 1974). This is an over-simplification. There is nothing to justify direct linkages between vicious trends today and the realities of youthful blitz experience. It did not work that way. Just as there are no signs that Father took against the garden after ninety wet nights in his Anderson so there is nothing convincing to suggest that his son thereby suffered claustrophobia or a fear of loud sounds by night.

Here of course we enter another difficult area for analysis despite recent improvements in social research methods. What impressed or affected the young, consciously or otherwise, inextricably ties up with intervening decades of falsification by memory. For adults, too, truth and fiction exaggeration, suppression, propaganda and national stereotyping have merged to form a luscious mishmash, where it can become impossible to distinguish the border-lines between fact and fantasy, dream and action – as with the extraordinary number of folk who now 'remember' being personally machine-gunned by German planes, one woman actually chased across open fields by a choosy Jerry pilot.

Re-adjustments of memory are normal, healthy and (after suffering) essential. If some who panicked in Plymouth have made it their finest hour, well, that is the strong meat of *Homo sapiens,* a carnivore insatiable in his determination to win.

The recall problem is, in the final count, crucial to any 'correct' description – and to a lesser extent interpretation – of this sort of mass social history. The more so in this decade, when it is normal in sociology or TV to interview people of all ages and to accept their versions of the past from memory. This interesting process is therefore the concern of a final chapter ...

12 Down Memory Lane

Remembering is not the re-excitation of innumerable fixed, lifeless and fragmentary traces. It is an imaginative reconstruction, or construction, built out of the relation of our attitude towards a whole active mass of organized past reactions or experience, and to a little outstanding detail which commonly appears in image or in language form. It is thus hardly ever really exact, even in the most rudimentary cases of rote recapitulation.

SIR FREDERICK BARTLETT
Remembering the Past

Those who cannot remember the past are condemned to repeat it.
GEORGE SANTAYANA

Bartlett, the great pioneer of memory research, as psychology professor at Cambridge, has proven in twenty ways that memory is self-deception. Yet a great part of what remains of the blitz in public minds is based on unchecked memory. For whereas in more ordinary times a personal reminiscence can commonly be checked against some sort of factual report, in war nearly everything published was coloured by the need to present good face and 'high morale' in the eyes of the enemy. The full truth of the blitz could seldom be told at the time; but what got into print then has largely determined the public record since.

For most surviving citizens the major effect has been (as often) in two opposite directions, both processes in 'reality obliteration': either to be unable to remember anything much (with no wish to do so), or, more usually, to see those nights as glorious. There is not much in between. But in between is where most of the unpublished evidence points – the evidence, that is, written down and filed immediately, without any intention of publication.

The process of public glossification in war – practised by Churchill down – is fundamental in assessing the values adopted and the conclusions offered in the preceding chapters. It makes it so difficult for anyone, including the reader, to sort out what is believable and what not. This writer indeed had the greatest difficulty in recognizing, in the seventies, what he had himself seen and reported in the forties.

To establish and if possible eliminate this fundamental difficulty, contacts have been resumed with observers of those days: war diarists have been asked to rewrite their experiences from memory, for comparison with their original accounts of the following morning; and in several cases it has been possible to follow up the release, through the Press Association, of old M-O and other blitz reports and to examine the public reactions to these today – in Coventry, Southampton, Manchester and Birmingham. The results are truly Bartlettian.

Let us first refresh the mind with the sort of 'loyal' incident which helped expunge the negative side of memory. It is 5 p.m. on Saturday 15 June 1940 in Putney High Street, south-west London. An observer notices some people gathered round a newspaper vendor on the pavement, who, in those economy days, wrote his own sales placards in black pencil. The paper on sale was the *Star*, mildly Liberal and now defunct. His result, in bold caps:

FRENCH TRAGEDY
IF NO ONE HELPS

Now two passing policemen cross the road and tell him to tear it down immediately. 'All right,' says the chap, 'only trying to make a living.' A young male onlooker: 'It's bloody near *sedition*.' France fell all the same.[1]

North of Putney lived the London observer who earlier described her first Hampstead bomb experience of 19 September 1940 (Chapter 5 (4)) as 'pure, flawless happiness'. Thirty-five years later she still looks back on it as a 'peak experience', recalling – along with very little of what happened – a sense of triumph and happiness, 'as if the whole thing was somehow a gigantic personal achievement'. A grandmother now, she compares it with 'the experience of having a baby'.

Further east, over in Stepney, the girl who was playing the

piano and missed Chamberlain's words or the first siren on 3 September 1939 (Chapter 3) completely transforms those events in memory. Then her mother burst in, shouting at her to stop playing, until Father took over, issuing dictatorial commands and unnecessary advice. Now she writes in recall: 'We were gathered in our little living room and it was very crowded, with the six or us (parents and four children) all together for once. But weren't there also visitors? I have the notion that this was a special kind of gathering; something a bit formal: aunts, uncles or neighbours, perhaps, all listening to the wireless, which, those days, was on almost all the time, in anticipation of more bad news.' A vigorous discussion followed Chamberlain's speech (she remembers). Then suddenly the siren. She is 'shaken to the roots' – in fact, she never heard it. Without a word on Father's key role, she rewrites:

Everyone was in a panic. Nobody knew what to do. Nobody, that is, except my mother who had read somewhere that the fumes of urine neutralized the effect of poison gas. To be honest, I'm not sure whether it was on that particular day or during the following week that she put her anti-gas plan into operation. But it makes a better finale to my recollections (and may well be accurate) if I relate that we were all solemnly made to pee into our chamber-pots, which were then placed beside every door in the house, and that, fortified by this safety device, our family was now ready to face the war.

For years she has been telling this story as a standard part of the war days, though it appears nowhere in her original documents which were extremely candid. Remember: she wrote it all down at the time too. Those who kept no records would normally distort even more.

After London, Coventry, where we tested another way, with whole-time observers sent there with M-O's special blitz unit. Three of those whose documentation on the spot we have were asked to recall what they could of this dramatic 1940 sequence. I personally have, for years, been telling of the heroic position there, beside the gutted cathedral, of a lone Salvation Army canteen-van. I have often said that for that reason I never ceased to admire and support the S A since (which is fairly true). Looking afresh at the original papers it is clear that its canteen did not arrive until the following morning, after a night of 'safety'.

George Hutchinson, who after the war became a journalist, biographer of Heath, head of Conservative Party Research and *Times* columnist (currently), at first could not remember anything about Coventry, except in the vaguest holocaustic terms. He was amazed when shown his own report on the Kenilworth trekking and could hardly accept he had stayed in the vicinity for more than a week.

Richard Fitter, now an eminent conservationist, ornithologist and botanist, could not remember having been to Coventry. He could hardly believe his eyes when shown his hand-written accounts of a long visit to the place including important conversations with leading high-ups to discuss after-measures.

On the civic side, too, Coventry memories were put to another sort of test in the seventies. Dr Paul Addison, lecturer in political science at Edinburgh University, drew on M-O and other intelligence sources while researching a piece for the *Sunday Times Magazine*, of 21 May 1972. In passing, he streamlined the Coventry information into: 'The first blitz on Coventry produced mass panic: thousands fled the town', and so on, ending his short paragraph by insisting that industrial production went up nevertheless. This passage sparked strong protest to the *Sunday Times* and locally. The Coventry *Evening Telegraph* ran columns of contradiction by interview and correspondence. One leading wartime alderman looked back on 'a triumph of improvisation'. Another was full of self-praise. The Deputy Chief Constable of the time recalled no kind of panic; instead 'people were very quiet'. A typical citizen remembered that 'the way the people were helping each other was wonderful'. The consensus in Coventry was for a magnificent, calm achievement. The word panic was a bit overdone in the original press story, to some extent obscuring the argument. This was corrected a few months later by another version, this time from the Press Association, quoting the M-O report on Coventration night more fully, with emphasis on the feelings of helplessness and initial fear rather than panic (see Appendix A). The local *Telegraph* repeated, verbatim, all the civic rebuttals of the previous year, plus one from the current Lord Mayor who said he'd been all through the bombing. 'Morale was exceptionally high,' he declared.

After Coventry, Southampton, with its longer series of double weekend raids from late November. Many years after the events, that gateway to the south was rudely re-opened to another double assault. On 14 February 1973 Southampton's blitz-veteran *Echo* ran a Press Association release banner-headlined: DID A BLITZ-HIT SOTON BELIEVE IT WAS FINISHED? Two columns of fine print told how 'investigators employed by the Mass-Observation Group' found local raid obsession 'becoming dangerously near neurosis'. Quoting from our 19 December 1940 report ('Aftermath of town blitzes') the *Echo* went on: 'Mass Observation concluded: "The human and morale problems of Southampton are being left to local resources, and local personalities, which are, in this case, inadequate".' Loud cries of protest assailed the paper's office. Rejections were respectively headlined: TOWNSFOLK SHOWED CONTEMPT FOR DEATH, BRAVE PEOPLE and THEY STAYED PUT (*Echo*, 19–23 February). No sooner had these angry voices died away than the town received a second blitz.

On 27 March, 1973 the *Echo* carried a huge front-page banner headline:

INDICTED:
SOTON LEADERS IN BLITZ
Mayor left on 3 p.m. train;
Town Clerk 'unsuitable'.

This was a release of the Hodsoll report. The *Echo* was very frank, even to the extent of printing 'the machine appeared to have come to a complete standstill'; the Town Clerk was 'entirely unsuitable' to his key post as ARP Controller; the rest of the city hall unfit to take over; 'the whole place riddled with intrigue'. Local reaction was even more violent this time, since the Wing-Commander's document had been much more personal than M-O's. Reached for comment, the old Town Clerk, now 85 and still living in the New Forest, replied simply: 'I never much cared for the job. I would have preferred to be in a battery of artillery.' No one had any immediate reply to Hodsoll's charge that 'the departments of the civic authority had lost touch with each other, and some of the Government (Whitehall) Departments had *no* contact with the local authority'. But next evening Alderman Sir James Matthews sternly denounced the whole report as 'a panicky statement'. The wartime Borough Treasurer, now 82, proclaimed that 'to say the civic machine . . . (came) to a complete standstill could *only* have been uttered by someone who never came to the town, and is an insult'. A spate of letters spoke out for the Sotonian GALLANTRY, GREAT SPIRIT, BOMBS AND BRAVERY, accused of MUCK-RAKING, all with heavy heroic captions.

More than a month later, the rumbles of this civic storm still rolled on. A wartime mayor, Rex Stranger, CBE (83) bravely flew in from his Jersey retirement to 'join battle against those who besmirch the proud name of Southampton'. He ended a powerful seven column counter-attack (LET'S FIGHT TO CLEAR OUR CITY'S NAME) by proposing a committee to answer in more detail: 'It is our duty to Southampton, our town, our city. If we do not do this it is quite certain the stigma will rest on our town for ever and no one will ever again be proud to accept the Freedom of Southampton.' (*Echo,* 10 May.) At the same time, a veteran wartime mass-observer, living in Southampton (her husband a native) found the glossified view prevalent in a rich collection of overheard conversations and interviews, but with notable exceptions. She added an important personal note suggesting how glossifying has gathered enchantment with the passing years and may not have been so strong sooner after victory:

No one has mentioned this, but I remember when I first came to Southampton in 1951, someone told me that 'After the big raid, when Southampton's High Street was left in ruins, and most of the centre of the town, the King and Queen came on a flying visit. As they went down the high street, people booed. But we were not booing the King and Queen. It was all the town's top brass who were with them. Everybody knew that they got out of Southampton every night and only came back to meet the King and Queen'.

Other publicity released from sample M-O reports on Bristol, Liverpool and Manchester produced similar results, nothing new, except that ordinary Mancunians seemed to show a low degree of interest in that piece of the past 'because the essence of the story was that morale in Manchester was low *because local leaders were inept*' (Dennis Forman, from Granada TV).

In Birmingham, full as ever of civic pride, the Press Association released the Home Security Research Department post-blitz study of Birmingham, including a reference to the possible 'dissolution of the civic entity', had the raids continued, and stressing grave local failures to learn from successive assaults. The *Birmingham Post,* 3 August 1974, was loud with rallying cries, one of which, ex-Lord Mayor Tiptaft on 'Nether Backwash', has already been quoted. Ex-Lord Mayor Mole ('who has recently taken the cloth') could only remember the city's people as 'splendid', ARP personnel 'magnificent'.[2]

It is hard to experience the double take of living through blitzes in 1940–41 and then living through the same scenes intellectually in 1972–5 without reaching again a conclusion which George Hutchinson quoted from this writer in a *Times* interview, 6 July 1973, when I ventured to assert that 'the only valid information for this *sort* of social history of war is that recorded at the time on the spot'. After two more years of study, this remains my view.[3]

This is the picture as clear as I can see it, returning to Britain and the documentation I then controlled, after a quarter of a century. And if, as Santayana observed, those who cannot remember the past are condemned to repeat it – well, I have done my small best to place memory as far as possible in focus. Few will dislike me the less for the painful effort. As Anthony Trollope's Alice Vavasour so drably observes: 'Bygones will not be bygones. It may be well for people to say so, but it is never true.' A century later, Simenon put it another way, commenting on Madame Marde's candid statement to the police when her husband had disappeared: 'Everything she had told the Superintendent was true, but sometimes nothing is *less* true than the truth.' It is the final dilemma in writing this sort of social history, with its unceasing conflict between the recollection and the record – along with the inadequacy of both.

POSTSCRIPT

Ultimate Uncertainties

The readiness to endure repeatedly even what was earlier hardly tolerable is one of man's greater strengths. In the narrow context of this study this closely relates to the human capacity to live through a blitz and come out ready for more of the same. Here one gets back to the question (raised in the beginning of Chapter 9) which can hardly help underlying much of the interest in this subject. Put crudely (1) *could* the Germans have won the war by bombing, with the equipment and skills they actually had, but by some different use of them? And (2) more broadly, if the answer is 'no' at (1), how is it for others? Are there other conditions where Douhet could be proved right? Can you bomb any organized people back into the Stone Age *and* thus force group defeat?

Let us again look for the last time at the blitzing of Britain to try to answer question (1). Battles like Coventry and Southampton have already been evaluated as temporary 'defeats' for Britons, in a narrow, local sense. Others followed. But Britain as a 'nation' kept going, only little affected industrially and militarily, certainly nowhere near the Douhet concept of the nation knocked-out by the overleaping of the armed forces and destruction of the civilian will.

While the rest is necessarily supposition, in my own opinion – based on observation then, contemplation since and analysis now – it *might* have been possible for the Luftwaffe alone to have brought the United Kingdom into at least some serious disunity, had they based their plans on a fuller appreciation of the humans whom they attacked. This is roughly what the Luftwaffe dropped in the blitz:

Places	*Period*	*Tons HE (approx.)*
London I	7 Sept. to 14 Nov. 1940	13,600
Provincial towns (as in Chap. VI–VIII)	14 Nov. to 16 May 1941	10,500
Other provincial towns	14 Nov. to 16 May 1941	1200
London II	14 Nov. to 16 May 1941	5200
Total:	London whole Blitz	18,800
	Provinces whole Blitz	11,700

Thus London, before Coventry, received well above the total for all other major attacks on Britain by piloted bombers. The unfailing regularity of these earlier attacks greatly raised the Londoners' ability to adjust and created the best organized centre in the country. Had the Führer not fallen for the obvious, how different the process might have been. Say Coventry had been fully attacked for three nights in a row, then Glasgow, then Belfast, with perhaps Southampton six nights in a row in between, or alternating to small evacuee towns outside (Fareham, Eastleigh, etc.). Say perhaps that some cities had been left deliberately free, with Lord Haw Haw broadcasting specious reasons for their omission. Say Lord Haw Haw and other tools of psychological warfare had been used in close liaison with all attacks – and before them. Say elements had been added to bombing with HE and incendiary: sound effects, smells, symbolic objects (or animals), a few armed parachutists on to associated evacuee and trekker countryside. And say London had been left entirely alone at first. Or, again, that certain week nights had been established as the ones to dread in certain places.

The angles are innumerable: to speculate along them is fairly futile now. Yet the crude thinking behind German bombing of Britain, and later vice versa, is unmistakably part, an essential part, of the attempt to answer our first question. The German mistake is the more extraordinary in that for four days, right at the start, they showed signs of being – in this sense – 'clever', when on 28–31 August 1940, they bombed Merseyside every night, before starting on London. But, after 455 tons, they gave that up – and did not return in force until 28 November (350 tons). Similarly the weekend concentration on Southampton, widely spaced early on, certainly had a special effect in establishing there a kind of (exaggerated) doom-night dread.

Much of this strategic crudity stems back, of course, to the presentation of the target as one vast, undefined, flat cake called 'morale'. And this in turn goes back to the whole contemptuous philosophy of Douhetism. But given more subtlety in attack, one has to wonder also, how Britons, under different (and especially provincial) pressures, would have been *able to show full defeat* had it proved possible to reduce a significant proportion of the people to that state. The rumour of martial law on Merseyside, which swept the kingdom, does indicate one bridge from fantasy to fact. But there is no evidence that real people may be brought to revolt or to react positively in anything like the way predicted as possible pre-war. Large numbers of individuals, especially housewives and mothers, got very fed up and weary, though not necessarily with the blitz any more than the discomforts, domestic difficulties, separations and uncertainties of war as a whole.

Only one element in the population even began to look as if it could conceivably organize any public opposition to the war: the Communist Party of Great Britain. But it was deeply uncertain of its position, as at first Russia and Germany were 'allies'. It became the most patriotic of groups once Germany attacked Russia.

Without some close organization, individuals, already 'on their own' in the post-blitz breakdowns, had neither mood nor mode for expressing defeat *positively*. A stricken city would scatter and survive. This did little lasting harm to the war system as a whole, but was enough to relieve the worst pressures on the people. Important here, too, was the island tradition, truly insular. The British lived surrounded by the sea, far more of a barrier then than nowadays. There did not seem to be the *lateral* enemy threat; no pressure ON THE GROUND, so ecologically unsettling to man, still a territorial animal. A hundred, even ten hairy nuns, dropped near the mouth of the Clyde, between the two big May attacks, plus parapadres to taste, might have put this pressure to the test. It is far easier for panic to start that way.

But there are other elements in keeping Britain or any equivalent nation stable. Air warfare is in its peculiar way particularly fair: it spreads its burdens over all alike. Privilege may help a bit, as in getting a lift out of Portsmouth, but by and large it looks a manifestly 'democratic' way of distributing destruction to rich and poor. Much was made, properly, of the direct hits on Buckingham Palace and the House of Commons. The leaders shared with the led; and this was something Churchill, especially, understood to be essential. If and as long as the leaders, ministers, generals are seen to be right there in the line of danger, any tendency to panic is greatly weakened, whatever the structure of society. It is probable that many more would have 'stayed put' in British provincial towns if local leadership had not become virtually invisible, in the days of doubt. Even so, one lesson of blitzing may be that many people take little notice of their leaders at the best or worst of times and can get on quite well left to themselves. Moreover, most ordinary people in those days were used to deprivation and lived every day with anxiety; so that side of the distress was not so much new as *additional*. The rich never understood this.

Are there then, conditions under which any people can be made to lose the war for their nation, as in our question (2)? If they *feel* defenceless, helpless, isolated, clearly YES: this was the 'lesson' too easily learned and then over-generalized from air attacks on unprotected villages in the Middle East and Abyssinia. Even then, once the groundlings are alerted there is scope for adaptation by taking to the hills or jungle, guerrilla warfare and so on.

So much hinges on the questions people have to ask themselves under such stress. Is it worth going on? Why not let them beat us, isn't anything better than THIS? These are questions so secret that, bound by wartime's compulsive loyalty – the tool to make civilized men break the commandments and be honoured for murder – they may only be whispered to wife or self. More than at any other time or on any other issue, private opinion is liable to be quite different from public. So, what people say in public may not be taken as a safe index of underlying 'defeatism'. Full verbal expression will only come in the final stages of collapse; collapse, that is, from the national, fighting point of view. But history, especially modern history, has come to rely largely on what people say and write. So has most of current sociology (in

Europe). Under the kind of situation now under discussion, almost all that ultimately matters is what people have planned or are prepared to *do*, personally. Will they run? More than that, though, will they defy legal controls, turn against each other; manifest the real panic which amounts to a breakdown not only in administration but in social ethics and normally acceptable individual behaviour. Murder, rape, mass looting, fighting among the people, occur at this level. Behaviour takes over from words.

It all sounds awfully possible, put like that; yet it never happened in Britain, even at the worst, unless for a matter of minutes in little corners of a night's disaster. No one stoned a commuter's car or punched a bobby. It doesn't seem to have happened in the other bombed countries of the Second World War, either. It is not just that British persistence equates with democracy. Fascist and communist societies put up with worse. Japanese dictatorship was influenced to surrender by Hiroshima; but American studies have shown that, even under those appalling conditions, civilian morale 'remained', shaken but solid. The Japanese were beaten militarily, before the bomb.

The North Vietnamese were blitzed on a scale which was never approached in the Second World War. But when the panic came, it was on a vast, hysterical scale right across the south, in face of a northern enemy which had not one bomber. And it was started by a fleeing army.

To pursue these questions further would be to take us further and further into uncertainty. We are already far enough. We have not touched on possible differences between 'morale' in one nation and another, though the writer's foreign reading makes him doubt if this factor matters. So long as other minimal conditions are met, as indicated above, it really seems that *Homo sapiens*, barely organized and maybe only superficially 'protected' can adapt to all forms of conventional bombing. The V–1 and V–2 later in the war introduced new needs for adjustment, needs that were also met – but that is another story.[1]

There is an adage popular among preachers: give a man a fish and feed him for a day; teach him to fish and you feed him for a life-time. In those nights of threatening death, no one fed the people fishes. So they went out and taught themselves to fish. It was enough to make near-nonsense of all the pre-war theories with which this book began. Those theories were based on contempt for the 'common man'. But even where his courage may have faltered, his common sense – and cunning too – kept body and mind alive, with that seemingly endless capacity for adaptation which, alas perhaps for the future of this earth, has made man paramount over all the other beasts.

The overall human effect of the blitz – rather like the whole operation itself – was predominantly negative. Put another way: though a great many learned more or less to live with heavy bombing, both admiring and despising themselves and their neighbours along the way, nevertheless few 'learned' anything much of lasting value. An inability to draw from it other than transitory lessons appears, indeed, for the vast majority to have been somehow inherent in the actual experience. Or, more exactly, in the experience followed by sublimation, obliteration, obfuscation; a form of recovery from shock by self-therapy as well as a mode of strengthening the will against more of the same to come.

Here one encounters little-understood areas of the mind, seldom tested *en masse* in 'ordinary times' and not even identified – let alone explored – at that time, when all was clouded in the blurred shadows of 'morale'.

However, we can here see that quick human adaptation *need* not mean learning more than personally, temporarily. Quick adjustment tends, too, to ignore 'inessentials' in favour of getting on with the main task, urgently. As we have tried, too, to show from the start, the human process is never simple, single and uniform. Nothing involving millions of individuals can be simple.

The atomic and the hydrogen bombs have threatened to cross all humane borderlines since 1941. But just as British bombing of Germany vastly increased the impact quantitatively without the expected corresponding effect of shattered morale, so the little existing evidence that has come out of Hiroshima and Nagasaki suggests that the effect there, though great in quantity, did not importantly differ from Plymouth and Coventry in quality. Fortunately, the American psychologist Irving Janis has analysed the available Japanese data (in his *Air War and Emotional Stress*).

Although both Japanese cities were caught completely by surprise, with no warning and no preparations, and although a very large proportion of the survivors initially believed their own homes or streets had been directly hit, only 5% subsequently reported feeling depression as a result. There is only one authoritative record of any sort of panic among a sizeable group of survivors, although 'a small proportion ... behaved impulsively and perhaps irrationally, for a brief period of time'. As Dr Janis concludes for both A-bombs: '... the meagre, fragmentary evidence available on overt behaviour does not provide substantial support for claims that panic, disorganized activity, or anti-social behaviour occurred on a mass scale.'

There was no sign of any 'sizeable frequency of inappropriate, negligible, or asocial behaviour'. Very large numbers in both cities *returned* to the devastated areas within 24 hours. So, to conclude: 'The A-bombing at Hiroshima and at Nagasaki produced no greater drop in morale than would be expected from a single raid of the type carried out during the Massed B-29 campaign against other Japanese cities.' There was also 'little sustained hostility against the United States', while *outside* the target areas 'the A-bombs had very little effect on the morale of the Japanese population'.[2]

'Morale', whatever it was, is and will be, can find a way to stay alive just as long as man can carry on living, even if as Kathleen Raine has observed: '... on the shores, after the tempest lie Fragments of past delight, and of past selves, Dead rooms and houses, with the strangled shells.'[3]

APPENDIX A

The Coventry Blitz

Some recommendations made by M-O arising out of their work in Coventry, relevant to future bombing of individual towns (from the end section of FR 495 of 18 November 1940, with minor deletions for brevity; cf. discussion in Chapter 6). The original report now begins . . .

Apart altogether from the special action points [sent in from Coventry on Saturday, November 16th] we venture to make certain suggestions:

1. In general, Coventry only confirmed much more detailed study done in the East End of London. It particularly confirmed our impression that much more powerful and imaginative organization is required to deal with the purely psychological and social effects of violent air attack. The apparatus for casualties, psychotic cases, fires, gas, debris and other obvious material consequences of the raid seem to be extremely efficient in nearly every respect. But the apparatus for dealing with homeless is not efficient. And the apparatus for dealing with the much larger group who are merely immensely upset is barely in existence. Yet this group, numbering perhaps two hundred thousand in Coventry, firstly determines the whole morale of the area, and all areas around which they spread as refugees; secondly, determines the vigour with which the rehabilitation of the town will be tackled; and thirdly, determines the production in those plants which can still work.
2. We believe that the whole tempo would have been altered in Coventry, if the authorities had expended 5% of their energy in considering the problems of those who had not been wounded, but had only had their windows broken and their ears bombarded by twelve hours of row. For instance, we would suggest as part of every Regional Organization:
    a. mobile canteens
    b. a special propaganda flying squad
    c. loudspeaker van, in the hands of skilful, unacademic commentators

d. special reserves of voluntary social workers
e. special facilities for getting newspapers delivered and sold in the streets

Many similar suggestions can be made. The main point is that such steps should be taken at once. Just as fire-engines moved into Coventry from as far as Wigan and police were drafted from all all over the Midlands (though there was not a great deal for some of them to do), so, as the second wave of reinforcements arriving with safety, and the dawn, should come a wave of social help, hot tea, information, personal organization and news, snatching people out of introversion, and linking them up again to the outside world, from which they are otherwise severed by lack of transport, lack of telephones, destruction of local press, destruction of electricity and thus many radio sets, dislocation of many other radios, of the windows, etc.

3. These and other simple steps had not apparently been planned for, and some of them had still not been taken by Saturday evening. It is largely a question of priority, and of developing the Civil Defence work which remains in embryo and largely unorganized. The social side of Civil Defence has been built up around the vague and voluntary idea of the Rest Centre. But the Rest Centre, as previously visualized, does not stand up to intensive air attack. Our studies in Coventry indicated that out of 15 Rest Centres there, 13 had been hit or damaged. This suggests a definite organization with Rest Centres on the periphery of every important town. If suitable places, e.g. in village schools, were chosen, one of the major deficiences of Rest Centres, the lack of first-class shelter, would no longer be a factor preventing people from using them, or preventing those who are using them from getting more upset.
4. Similarly, there could have been much better organization of transport in Coventry. The debris was cleared from the streets with great rapidity and by early evening it was possible to drive nearly everywhere in the town. Yet there were virtually no buses running, and great dislocation in consequence. At the same time, a number of buses used to transport police, and many army lorries, to transport troops, were idle all day. It should have been quite easy for these to have been loaned to the Civil Defence, for at least several hours during the day. Incidentally, one of the few commodities available in the town was petrol! One of the biggest tangible effects of the raid on Coventry was the astonishing number of houses with no windows. Supplies of suitable boarding material were soon exhausted and anyway are beyond the means

of the poorer sections of the community. A few lorry loads of Essex boarding would have enabled people to black out the window of their main living room.

5. On the human side, we stress the undesirability of the extremely exaggerated accounts of 'marvellous courage' put out in the press. These are out of key with real feelings in Coventry, the courage of which is enough not to require being turned into a miracle. The *Daily Mail*'s ballyhooing of morale was particularly commented upon. The need for immediate factual coverage both of the damage and of the people, is just another part of a new development suggested for Civil Defence. Authoritative information of this sort ought to be available within a few hours and rapidly followed up by documentary camera, skilled commentator, a really good and balanced press story from the Regional Commissioner's office. Exaggerated treatment makes people suspicious and encourages few.

These were only the first of many such suggestions, the effect of which was to arouse anger rather than action among the relevant recipients. In his Official War History, Terence O'Brien concluded, mildly indeed: 'One of the main lessons of the Coventry raid was the need for much closer cooperation between services for which the Ministry of Home Security was not directly responsible, e.g. public health, food, industrial production, information services, housing and repairs to gas and water services.' The next harsh attack on Coventry came on 8 April 1941 (237 bombers); and much was found wanting again. (FR 790.)

APPENDIX B

Tiptaft of Birmingham

It is by no means the intention here to ridicule men of the quality of Alderman (and later Lord Mayor) Tiptaft (see Chapters 8.2 and 10, note 1). These have made a major contribution to local government in Britain, in a classic provincial tradition with some roots even older than medieval. A fair example of the powerful background behind such civic aldermanhood (which worked well in pre-war days) is provided by Mr Tiptaft's own serial contributions to the columns of the wartime *Birmingham Weekly Post*, where he wrote in neo-biblical style, THE BOOK OF A.R.P. Here is a pre-blitz effort, in the issue of 5 July 1940 (slightly cut in length):

The Book of A.R.P.
Of the First Raid
And how the people did behave
By Norman Tiptaft

Now in the days when men called the war a 'phoney' war and the Volunteers became tired of having little to do, or as they said in the manner of speech common among the people, 'fed up', many things seemed to go ill in the ARP. For the Wardens did argue that they were greater than those at the Stations of Auxiliary Fire; and they at the Stations of Auxiliary Fire did ask among themselves the value of those at the Posts of First Aid.

And although Jonand did urge all men in ARP to work together in peace and harmony, yet it was very difficult to accomplish.

Moreover, those Volunteers who came to the Stations and Posts for the whole of their labour complained because the time of it was lengthened to seventy-two hours a week.

And many made an excuse to leave Arp since they said 'this is wrong with it' and 'that is wrong with it' and many who had borne a good heart become exceedingly discouraged.

And the spies of Hit formed a fifth column, and the Fifth Column spread many rumours and lying tales, and did strive to stir up strife in Arp.

And their leaders sent a message to Hit saying: 'Send now aeroplanes against Eng, for the people are weary, and when ye send the aeroplanes and bomb the country, they of Arp will not be able to protect the people, and many shall be slain, and the men of Eng being defeated in their own homes will surrender'.

So Hit sent his aeroplanes against the land of Eng and they dropped bombs and fire.

They Did Go To It

But it happened not at all as the spies of Hit had told him.

For they of the Posts of First Aid, who had become weary, when the first raid started said: 'Surely now is our training of use, and we go to practise it.'

And they of the Wardens Service, whose duty it was to keep watch, and who had become tired of keeping watch, said 'Now is there something to look for, wherefore we go to see it.'

And they of the Control Centres, who had sat waiting for messages to come and had received nothing worthwhile to report, said 'Lo! there will be much doing at the Control Centres tonight. Let us go there.'

And every man and every woman who had a task to do for Arp did suddenly go to it.

Every Man at His Post

And when the raiders came, they who did conduct the people to shelter were there, and the people did take cover, and there were few who were hurt.

And the few who were hurt were gathered quickly into the Posts of First Aid, and their wounds bound up.

After the first raid was over, those who were in Arp were more than ever before, and eager to do their work.

And the things that the men said they could not do, they did; and the things of which they had complained they forgot.

And when Hit heard of these things, he said: 'Now is everything gone wrong. For we planned to do the things that they did not expect, and lo! the boot is on the other leg, for it is they who have done things that the spies told me they would not do.

'Truly, ye cannot depend on the men of Eng, for if we cannot break their Home Front, then we shall never break their warriors on land, or on the sea, or in the air.' And he said to his servant: 'Bring me a large portion of aspirin, for my head doth ache.'

And his servant brought it saying: 'If they of Arp, which is but for defence, give thee such a headache, what shall come upon thee when the warriors of Eng shall attack us?'

And Hit answered him nothing, for indeed he knew not what should be before him.

Thanks for drawing attention to the above are due to Susan Briggs, whose own album of the war, *Smiling Through* (Weidenfeld and Nicolson, 1975) deals lightly but sensitively with the blitz. See also Anthony Sutcliffe and Roger Smith, *Birmingham 1939–70* (Oxford University Press, 1974) for other Tiptaftia.

APPENDIX C

List of Tables in Text

[Those asterisked relate to several chapters in addition to the one where the table is placed; those asterisked and italicized may be useful for reference all through the main text.]

REFERENCES

As stated in the Preface, this book is based largely on unpublished material in the Mass-Observation Archive at the University of Sussex. But of course other work has been frequently consulted. Where these refer to particular or local themes, they are cited in full in the notes to each chapter. In addition, the following have been of great help over the whole field, and are often cited simply by the author's name followed by the relevant page citation (either in the main text or the associated notes):

CALDER, Angus. 1971. *The People's War*. London, Panther. (2nd paperback edition of 1969 Cape original.)

CALVOCORESSI, Peter and WINT, Guy. 1972. *Total War*. London, Allen Lane.

COLLIER, Basil. 1957. *The Defence of the United Kingdom*. London, HMSO.

NICOLSON, Harold. 1971. *Diaries and Letters, 1939–45*. (Ed. Nigel Nicolson.) London, Fontana (paperback edition of 1967 Collins original).

O'BRIEN, Terence H. 1955. *Civil Defence*. London, HMSO.

TITMUSS, Richard M. 1950. *Problems of Social Policy*. London, HMSO.

Mass-Observation's own publications are also cited in brief, as they are listed fully in the first of the notes (Preface, note 1).

NOTES

PREFACE

1. As the surviving active co-founder of Mass-Observation (the others took no part in the blitz or related war work) who has retained an active interest in the system's papers from 1937 on, these were passed back to me by agreement when I retired from the Colonial Civil Service in 1967. Then, in disorder, they were deposited at the University of Sussex for a trial period, with a generous grant from the Leverhulme Foundation Trust. In 1975 I deeded them to the University as a charitable Trust (first trustees: Asa Briggs as Vice-Chancellor, Professor of Anthropology David Pocock [who has made a major contribution], James Fulton CMG [ex-Foreign Office] and Henry Novy [one of our best war-time observers, now running an international operations research business]). Thanks are due to successive Sussex University Librarians and Library staff (especially Mr A. L. Pollard) over these years. From mid-1975 the Archive became a separate unit in the new Arts Building D, with myself as Visiting Professor and Honorary Director (pro tem. age 64), David Mellor (Executive Director), Dorothy Wainwright (Secretary).

   A general account of the Archive's contents was given to the Association of Special Libraries and Information Bureaux in a paper ('The Mass-Observation Archive') published in *Aslib Proceedings*, 1971, pp. 397–412, available on application. See also, recently: Postscript to Leonard Mosley's *London Under Fire*, Pan Paperback, 1971, pp. 447–51; Angus Calder's *The People's War*, Panther Paperback (second edition), 1971; Introduction to photo-album *Britain in the 30s: Worktown by Camera*, by Humphrey Spender and T. H., Royal College of Arts (Unicorn Press), 1975; and George Hutchinson in *The Times* 6.7.73. (Cf. Chapter V, note 2 below.)

   The following is a list of main earlier M-O books by the author and/or other mass observers published over the years.
   1937 *May 12th, 1937* [the Coronation]. Faber and Faber.

1938 *First Year's Work*, 1937–8. Lindsay Drummond.
1939 *Britain*. Penguin Special.
1940 *War Begins at Home*. Chatto and Windus.
1941 *Clothes Rationing*. Advertising Service Guild.
1941 *Home Propaganda*. Advertising Service Guild.
1941 *A Savings Survey* (*Working Class*). Privately printed for circulation.
1942 *People in Production*. Advertising Service Guild (John Murray).
1943 *War Factory*. Victor Gollancz (2 editions).
1943 *The Pub and the People*. Victor Gollancz (3 editions).
1944 *The Journey Home*. Advertising Service Guild (John Murray).
1945 *Britain and her Birthrate*. Advertising Service Guild (John Murray).
1947 *Puzzled People*. Victor Gollancz.
1947 *Brown's and Chester*. Lindsay Drummond.
1947 *Exmoor Village*. Harrap.
1948 *Juvenile Delinquency*. Falcon Press.
1949 *The Press and its Readers*. Art and Technics Ltd.
1949 *On Sunday*. Naldrett Press (illus. by Ronald Searle).
1949 *At the Doctor's*. Naldrett Press (illus. by Ronald Searle).
1949 *People and Paint*. I C I publication.
1950 *The Voters' Choice*. Art and Technics Ltd.
1961 *Britain Revisited*. Gollancz.
1966 *Long to Reign over Us*. Kimber.
1971 *The Pub and the People* (new revised edition). Seven Dials Press.
1975 *Britain in the Thirties*. Unicorn Press (Royal College of Arts).

Many books have discussed M-O at various levels. The following contain interesting passages on M-O's (alleged) methods:

The Confidential Agent by Graham Greene, Heinemann, 1939.
The Divine Flame by Sir Alister Hardy. Collins, 1966.
Into this Dangerous World by Woodrow Wyatt. Harrap, 1952.
Indigo Days by Julian Trevelyan. McGibbon and Kee, 1957.
The Thirties by Julian Symons. Cresset Press, 1960.
The Thirties by Malcolm Muggeridge. Collins, 1969.
The Age of Illusion by Robert Blythe. Hamish Hamilton, 1963.
The People's War by Angus Calder. Panther (preface), 1971.
Kingsley by C. H. Rolph. Victor Gollancz, 1973.
The Land Unknown by Kathleen Raine. Hamish Hamilton, 1975.

Pre-Archive, and from 1943, M-O's activities were directed by Bob Willcock and then, for many years, by Leonard England and Mollie Tarrant, all war-time Observers. M-O now continues separately as a market research business, Mass-Observation (UK) Ltd, 229 Shepherd's Bush Road, London, W6, with Managing-Director John Parfitt, who is also Chairman of the Youth Hostels Association.

2. We follow Basil Collier's official *The Defence of the United Kingdom*, HMSO, 1957, for the now widely used set of figures for raids, German planes and bombs involved over Britain. These, based on German sources, are adequate for the present comparative purposes. The actual impact of High Explosive on the ground was often considerably less, but the accurate information on this point is only available on a patchy local basis – as discussed at various points in the following text. Frequent reference will be made by name to Collier and the other main relevant official war historians as formal background; their fuller titles are listed at the end of the preface; cf. chapter VII(1), note 15.
3. As M-O's system of war diarists and voluntary informants (the 'national panel') worked on a basis of strict anonymity, thus encouraging complete candour, the material cited from these sources can only be checked back (by permission) in the Archive itself. Special help in sorting this massive documentation – not yet indexed – was given by Mr H. D. Willcock, OBE, a main report and book writer for M-O and later in peacetime for the government's Social Survey, and Mrs Sallie Rée at Falmer. Other M-O report writers of the early war period were the late James Fisher, Richard Fitter, John Ferraby and Alan Hodge (now editor of *History Today*).

   The more general text here relies considerably on reports written up in the office at Ladbroke Road, W11, notably for Home Intelligence in the Ministry of Information, Naval Intelligence and the Advertising Service Guild (on propaganda aspects). These are now in typed form, numbering tens of thousands of pages, and hereinafter referred to as FR (File Reports) with appropriate sequence number (FR 1272) as in the file cabinets. Other untyped raw materials (RM) from many sources are used where suitable in support, usually with a box and file (F) number if the relevant material is *concentrated* in any one section.

   Major assistance in preparing and typing drafts has been received from Mrs Ann Ginnings and Mrs Dorothy Wainwright at Brighton; Mrs Michael Novy, Mrs Diana Forrest and Mrs Jennie McKoen in Brussels. Other valued help has been generously given by Mrs Naomi (Lady) Mitchison, Robin (Lord) Maugham, Lord Kaldor, Dr David Lumsden at the Stockholm Institute for Peace, Professor David Martin at LSE, Dr John Maynard at Guy's

Hospital, Paul Rotha, Michael Shanks, A. A. H. Knightsbridge of the Public Record Office, C. J. Child and Miss H. Merrifield at the Cabinet Offices (Historical Section), Lord Horder, Herman Grisewood, John Noone (British Council), Tom Hopkinson (wartime editor of *Picture Post*), Professor Roger Eddison FRS, Professor Christopher Cornford and Humphrey Spender at the Royal College of Arts, Tom Driberg, John and Miranda Raynor, as well as those mentioned on specific matters in what follows. The debt to Celia Fremlin also needs re-emphasis here (see last page of preface), while David Mellor has helped in many ways along the line from Sussex, as has Lt-Colonel Leo van Zwynsvoorde at the École de Guerre, Brussels.

1 THE EXPECTED HOLOCAUST

1. Background in Col. Raymond Sleeper, *Air University Quarterly*, 1951. 1:5 – a journal (later called *Air University Review*) full of valuable perspectives (e.g. Col. A. P. Sights on 'Strategic Bombing and Changing Times', 1972, pp. 14–26). Wing-Commander H. R. Allen, *The Legacy of Trenchard*, Cassell, 1972, p. 17. A. J. P. Taylor, *The First World War*, Penguin, 1972, p. 228. Mr Pemberton-Billing, aeroplane manufacturer, who played a prominent part in the 1917 parliamentary furore, appeared in a large Rolls-Royce as a peripatetic 'Reprisals Candidate' in the 1940–41 by-elections (cf. chapter 11, 6). On Churchill's early reactions see also Calvocoressi and Wint, p. 491 (see reference list after preface), and Robert Rhodes James, *Churchill: a study in failure, 1900–39*, Penguin, 1973. A good study of the bombing in the First World War is in O'Brien, op. cit., pp. 7–12 (see preface references).

   On reasons for distrust of Churchill see, for instance, succinctly, Christopher Farman, *May 1926, The General Strike*, Panther, 1974, p. 35; and Rhodes James, op. cit., also our chapter 11, 5 and elsewhere, passim. See also chapter 11, note 13.

   A clear picture of how a very sensitive yet 'universal' person could be affected in 1917–18 is given in Professor Quentin Bell's fine biography of his aunt, *Virginia Woolf*, Hogarth, ch. II, pp. 52, 217, etc.; the effect of bomb fear on her huddled in a basement 'may have been therapeutic'. From the time when she came under German attack, all talk of suicide (previously an obsessive theme) ceased. When her London home was blasted she felt: 'an odd, unaccountable sense of exhilaration'. Mrs Woolf reacted then like many others later (see chapter 5, 4).
2. On Douhet no one analysis really satisfies. See Col. Vauthier, *Le*

Doctrine de Guerre du General Douhet, Berger-Levrault, 1935. As early as 1934, the Polish General Sikorski (later Polish leader in exile) had written in his own language a powerful critique of Douhet, translated into French (with foreword by Marshal Pétain!) in 1935, but not into English until 1943 (as *Modern Warfare*, Hutchinson). At p. 116 he concludes that 'a nation whose *morale* is solid will not declare itself beaten' because of air attack. His views, ignored in Britain pre-war, affected some of Field-Marshal Goering's key Luftwaffe advisers, though not fully Goering himself. See also Calvocoressi and Wint, chapter 23; H. R. Allen, *Who won the Battle of Britain*, Barker, 1974, p. 8; Collier, op. cit., p. 2. Also H. A. Jones, *The War in the Air*, HMSO, 1935, ch. v: appendices 6 and 7; O'Brien, op. cit., p. 17; and Goering. L. Mosley, *The Reich Marshal*, Doubleday (NY), 1974, p. 259; cf. p. 243.

3. (Lord) Portal, then Squadron-Leader, took an active part as early as a 1923 high-level meeting on aircraft policy (chaired by Trenchard) while he was attached to the Director of Operations, Air Ministry (minutes in C. Webster and N. Frankland, *The Strategic Air Offensive against Germany*, a monumental study; HMSO, 1961, ch. v, pp. 62–70, cf. pp. 71–83). For Trenchard's life and temper, see especially Andrew Boyle's *The Legacy of Trenchard*, Collins, 1962 to be read beside Wing-Commander Allen's *Who won the Battle of Britain* (already cited; especially his chapters 4 and 5). One of the least partial background accounts of this era is Harold (now Lord) Balfour's *Wings over Westminster*, Hutchinson, 1973, including on Portal his pp. 101, 107, etc.
4. The crucial 1941 Trenchard memo and reactions are fully given in Webster and Frankland (note 3), op. cit.,ch. v, pp. 194–200; cf. the air staff directions at *idem*: p. 135 on. On 9 July 1941 the RAF learned, after the British blitz was over, that a 'comprehensive review of the enemy's political, economic and military situation discloses that the weakest points in his armour lie in the morale of the civilian populations', leading to 'an outline plan of attack on German transportation and morale'. This 'disclosure', like Trenchard's forceful arguments of three weeks before, was not based on any acceptable objective criteria (see further at chapter 10).
5. Harris/Douhet quote from Mark Arnold-Forster's lucid *The World at War*, Collins, 1973, p. 274. For enemy countries the information on so-called strategic bombing (often a euphemism for indiscriminate civilian assault) is largely based on post-war interviewing and statistical studies contained in the British and United States Surveys of the Offensive in Germany: cf. Frankland and Webster (as at

note 3), op. cit., ch. v, pp. 40–60 and pp. 465–525. Photocopies of the main US research are now in the Archive, thanks to Jo Sinclair and James Fulton.

6. Churchill in *Hansard*, as for all other House of Commons items hereinafter quoted. Lindemann (Cherwell) letter in *The Times*, 8.8.34. The influence of Cherwell on Churchill has been much discussed, notably by Roy Harrod, *The Prof.: a Personal Memoir*, Macmillan, 1959 and Lord Birkenhead, *The Prof. in Two Worlds*, Collins, 1961. The latter emphasizes (pp. 98–9) him as 'capable of taking a spiteful and petty revenge' – with his top hate 'the Hun'. Sir Roy, Oxford colleague, describes him as 'revengeful' and 'out of touch with the course of contemporary thought' (p. 87). A fair estimate is given in Christopher Tunney's useful *A Biographical Dictionary of World War II*, Dent, 1972, p. 30, where the Cherwell entry mentions a 1942 minute to Churchill estimating bombs would smash German morale into defeat. His influence is considered by some to have prolonged the war by up to 12 months. See also Rhodes James (note above).
7. During the full 1939–45 war span, under 90,000 persons (5 per cent of expectation) were hospitalized in Britain as the result of any and all forms of enemy attack (Collier, op. cit., p. 528 and O'Brien, op. cit., p. 560; both give 86,182).
8. About 180 tons were actually dropped from the declaration of war to the end of 1939. A year later, at the accepted start of the London blitz, 33 tons of HE were sent down at first, with a peak of 538 tons on the 15 October. 200 tons was a 24-hour average at this period – Coventry, on 14 November, received (theoretically) 503 tons. The statistical weakness of the basic pre-war calculations (the 50 casualty per ton factor) is aptly demolished in a terse footnote by Titmuss, p. 12, which deserves immortalization.
9. Negley Farson, *Bomber's Moon*, Gollancz, 1941: John Langdon-Davies, *Air Raid*, Routledge, 1938: *The Nature of the Air Threat* (pamphlet from AR Defense League), 1939; J. B. S. Haldane, *ARP* (Chapman and Hall and Gollancz), 1938; and Melitta Schmideberg in *Int. Journal of Psychoanalysis* (London), 1942, pp. 146–76, a sober survey of a kind very rare in the contemporary specialist literature. Haldane's is perhaps the most detailed and (from a scientist) revealing illustration of how completely the holocaust/panic concept had infiltrated western democratic societies, from far right to far left. But the same reasoning on the left produced more positive conclusions, as 'even a million Air Wardens' would not be able to prevent mass panic, if not massacre, the only solution was to build *deep*, bomb-proof shelters under

every city. See also D. Cameron Watt, *Too Serious a Business*, Temple Smith, 1975, chapter 5 ('Reluctant Warriors').

10. Some of M-O's earlier studies in the thirties showed a truly monumental public ignorance of and lack of interest in international affairs. Many did not know who was their Prime Minister or local MP. These 'revelations' helped attract wide interest in our work. See, for instance, M-O's *Britain* (1939) and *War begins at Home* (1940), as cited in preface note (1), above.

2 AIR-RAID PRECAUTIONS (AND PEOPLE)

1. W. S. Churchill, *The Gathering Storm* (vol. I of his *The Second World War*), MacMillan, 1948, p. 115. Thanks to Stephen Spender, old friend, for permission to use part of his 1942 poem, *Fireman*, originally published in Ingram (note 6 below).
2. The shaky relationship between the centre and the provinces becomes a main theme in chapters 6 to 8 following, analysed in chapter 10. The Regional Commissioners (two for London), who assumed office in late August 1939, included three politicians previously in high office, three retired civil servants, four industrialists, an admiral, a general and a Cambridge don (who afterwards became a politician). An RC's office proved visibly more effective when it was at the bombed spot (e.g. London, Bristol) than far away (in Reading cf. Southampton and Portsmouth, Bristol cf. Plymouth and so on). In Scotland the sub-division into District Commissions was much more realistic.

   O'Brien has good general material (especially his chapter 5) on RC and ARP systems generally, though nothing on the receiving end; he confesses he only visited two places outside London in preparing his official history.
3. The statistics of the war's inter-service defence measures and their results are fully discussed in Collier (see preface reference list).
4. Titmuss, op. cit., p. 102; his chapters 3, 7, 8, 10, 18 and 21 fully deal with evacuation arrangements, and the Archive has much on how these worked in human terms. The poverty of official data is indeed astonishing (cf. Titmuss, op. cit., p. 356).
5. 'Trekking' is discussed fully in our chapters 6–8: cf. Titmuss, pp. 272, 306.
6. Archive FR 296 of 25.7.40; also FR 536, 578, 597; RM (Raw Material) Box 38/F 3678, 40/F 3424 and 80/F 4170. On the rich literature of the fire-war the most interesting is the anthology *Fire and Water*, edited by H. S. Ingram for the National Fire Service, Lindsay Drummond, 1942.

7. The growth of this situation pre-war is well traced in Titmuss' chapter 4, which may be compared with M-O's observations of how this worked out in the blitzes to be described (see especially Plymouth at chapter 7, 5).
8. The negligible psychiatric results of bombing are examined further in chapter 11, 2, and the literature reviewed there. The sanest post-war post-mortem is by Dr Irving Janis (for the Rand Corporation), *Air War and Emotional Stress*, McGraw Hill (New York), 1951, especially his chapter 5.
9. This table is based primarily on O'Brien, op. cit., pp. 229–315 and pp. 685–6, elaborated by our own observations. There were variants in Scotland.
10. F 716 of 27.5.41, one of the series of special reports on blitz and post-blitz; see further later.

3 THE FIRST SIREN

1. Extract from one of several blitz poems from Naomi Mitchison, observer and war-diarist (named by permission).
2. The material in this chapter comes especially from war diaries for 3.9.39, plus observer casual reports and raw material for that day. See also M-O's *War begins at Home*, pp. 43–9; *Britain*, chapter 2; FR 371.

4 THE FIRST BOMBS

1. Every first enemy plane, bomb or A A gunfire attracted lavish comment in war diaries and collected overheards. Also for 'phoney war' phase FR 157, 164, 167, 170, 172, 182, 191, 212, 217; and a re-analysis of RM (Raw Material) by Celia Fremlin, April 1973.
2. Summer 1940 covered especially by FR 261, 266, 285, 315, 355, 366 a–c; and village intensive report series, as well as diaries. Background of organization in O'Brien and Collier. For some censorship effects on bias or uniformity in published reports see especially *Blue Pencil Admiral* (1947), by the chief wartime press censor, Admiral G. Thomson.
3. RM 293/37, F 3495 of 3475; FR 364, 370, 382, 585, etc. Information courtesy Harrow Town Clerk (Mr Stanley Lancaster). Also Calvocoressi and Wint, op. cit., chapter 23; Calder, op. cit., pp. 176–8.
4. The statements by Goering and Gulland are from Leonard Mosley, *The Reich Marshal*, Doubleday, 1974, pp. 257–9. See also a

'biblical' response to the first Birmingham bomb, July 1940, at appendix B.

5 THE LONDON BLITZ

1. This chapter is entirely based on previously unpublished M-O data, especially very detailed reports by both voluntary and whole-time observers living in London. Celia Fremlin did the major work of assembling this information and preparing a full first version (1973). Other published treatments based partly or largely on our files, before the Archive was organized and all the papers were available, are by Angus Calder, Constantine FitzGibbon, Leonard Mosley and Richard Collier (see further below at note 9 and chapter 8 (6), note 9).
2. For another mass funeral see Portsmouth in chapter 7, 2.
3. See the present memory of this new grandmother, chapter 12.
4. FR 408 a–d is a 60-page first study of 'Human Adjustment in Air Raids' (September 1940); also FR 436; RM F 3442, 3494/7 etc. Collier: 489 on AA guns.
5. Unknown to us at the time, Dr E. J. Lindgren was making a separate, shorter study of sleep in Shoreditch (East End), 20 September, resulting in: 33% less than 3 hours, 30% 3–6 hours, 27% over 6 hours. Background information on normal sleep in the thirties and forties was poor; cf. later work by G. S. Tune, *British Medical Journal*, May 1968, p. 269; J. P. Masterson, *Lancet*, 1965, p. 41; also Ian Oswald, *Sleep*, Penguin, 1966; W. P. Colquhoun, *Aspects of Human Efficiency: Diurnal Rhythm and the Loss of Sleep*, English University Press, 1972.
6. FR 464 (prepared for Zuckermann), pp. 489, 1423 on ear-plugs; cf. RM 3371, 3474, etc; *Lancet*, 1940 (October), p. 459 (editorial); *British Medical Journal*, 1940 (November), p. 396. On wider issues Erich Fromm's 1942 *The Fear of Freedom* and recently Ernest Becker, *The Denial of Death*, Free Press (NY), 1973; cf. Calder, op. cit., p. 196 and Titmuss, op. cit., p. 384.
7. Many raw reports plus an inadequate M-O summary, 'The Tube Dwellers' in *Saturday Book*, 1943; see also Ritchie Calder's powerful polemical analyses of breakdowns in *Carry on London* and *The Lesson of London* (both 1940), and Charles Graves, *London Transport carried on*, LPTB official story, 1947. Also T.H. in *New Statesman*, Sept. 1940 and Susan Briggs, *Smiling Through*, Weidenfeld, 1975, p. 103.
8. Titmuss, op. cit., p. 251; O'Brien, op. cit., p. 392; cf. Calder, op. cit. p. 216.

9. Calder has a useful bibliography of London blitz literature, and his own book (see preface references) covers it in summary form, supported with some M-O material. The two books of excellent wartime journalism by his father (now Lord Ritchie-Calder) are referred to at note 7 above; see also his article in *New Statesman*, 21.9.40. Leonard Mosley's *London under Fire*, Pan, 1971 (originally published by Weidenfeld and Nicolson as *Backs to the Walls*) covers the whole war in London, heavily using M-O diarists. Constantine FitzGibbon's more imaginative *The Blitz*, Wingate, 1957 (re-issued by MacDonald in 1970), despite its title deals only with London, and similarly acknowledges debt to M-O (now Archive) sources. A stirring though naturally hyper-patriotic book of strong local quality is *Cockney Campaign* by the blitz-time Mayor of Stepney, Frank Lewey (Stanley Paul, 1944). Most of the rest is glorification and glossification, at its worst in Basil Woon's *Hell came to London*, where the text is interspersed with italicized paragraphs thus:

   Whoo-parrmph-woosh – – – –

   the noises of the bombs which cascade around the ebullient author. See also chapter 8, 6 for the final phases of bombing London.

6 'COVENTRATION'

1. Hilde Marchant, *Women and Children Last*, Gollancz, 1941, chapter IX. TH text from BBC (microfilm) Archives, 16 November 1940, thanks to Roy Plomley and Pat Daniels. The latter text also referred to the lack of demand for reprisals which we found in Coventry.
2. Main data summarized in FR 495 a–b, 497; cf. RM Box 38/80, F 4170 etc. Pre-blitz F 3436 includes a fascinating account of M-O co-founder Charles Madge and colleagues in a June 1940 Coventry 'alert'. Also re-analysis of war diaries relevant to Coventry's radial effects by H. Willcock 8.2.73 (excluded here for reasons of space). Calder, op. cit., pp. 234–7 draws on FR 495. One of the Coventry observers was George Hutchinson (see chapter 12); another Richard Fitter (appendix A and 12).
3. MHS Civil Defence Research Committee, RC 107 by Professor J. D. Bernal and F. Garwood. This study has often been cited out of context to show the cleverness of 'operation research' predictions. On analysis, it is in some important respects almost as misleading as the pre-war predictions which also perhaps unconsciously affected the brilliant but slightly 'inhuman' approach of the two scientists (cf. Titmuss, op. cit., p. 329).

4. Churchill's *Second World War*, vol. II, p. 79; see further at chapter 8 (2) on Birmingham. A. J. P. Taylor, *English History 1914–45*, 1965, p. 502; cf. Collier, op. cit., p. 265, for other factory details, as also for Coventry's air defences. It is significant that P. Inman's official *Labour in the Munitions Industries* (HMSO 1957), ignores air-raid effects except, *passim*, in the 'southern dockyards' (p. 99) – cf. further at Portsmouth in chapter 7 below and again at Birmingham in 8; and in 11 generally.

   H. M. D. Parker's official *Manpower* (in war) (HMSO 1957) makes it abundantly clear that other factors presented much greater difficulties in planning the war economy. He has only one significant reference, at p. 443, where in writing about hours of work he mentions 'Air raids were an additional complication', leading to rearranged times so that workers could get home before dark (see further at chapter 11, 1).
5. See appendix A for part of FR 495 and reference to FR 790. For Coventry's 'official' reaction to observations on their blitz as given here, see further in chapter 12. It must be added with emphasis that M-O, like 'everyone else' (apparently) had by this time become so conditioned to the London blitz that Coventry took us by surprise too.

7 THE SOUTHERN PORTS

(*1*) *Southampton, 'The English Gateway'*

1. '*The English Gateway*' is the sub-title of the demi-official war history of Southampton (Knowles, 1951) referred to and discussed later in this chapter.
2. Archive RM Box 80/F 4170; FR 516, etc.
3. Albert Speer, *Inside the Third Reich*, Sphere Books, 1971: pp. 392, 450, 468, 476, 723, 731, 741, etc.
4. Background data from M-O files; Bernard Knowles, *Southampton: The English Gateway*, 1951, p. 148 and appendix A; Paul Nicholson, dissertation for University of Hull, 1973; information from local authorities, 1974; cf. Calder, op. cit., p. 183.
5. Nicholson, op. cit., 1973; cf. Calder, op. cit., p. 235.
6. J. B. Priestley, *The English Journey*, 1934, p. 397 etc.; Nikolaus Pevsner, *The Buildings of England: Hampshire and the Isle of Wight*, Oxford, 1967, p. 526; Archive FR 516 a–d and following.
7. This and following contemporary remarks collected during current (1973–4) continuing work in Southampton by wartime investigator Marian Sullivan.

8. Diocesan views in Knowles, op. cit. 1951; Nicholson's dissertation, op. cit., 1973; cf. Calder, op. cit., 1971, p. 251; and the Archbishop's biography (Cyril Smith).
9. Hodsoll's wider role has been indicated in chapter 2, Morrison's anger is illustrated in 10. The Hodsoll report is now available in PRO files; cf. HO 199/322 9HSI 125/3. This material was first recognized as of wide public interest by Mr Tom Corby, a brilliant young journalist on the Press Association's Special Reporting Unit (directed by Mr L. B. Cubey), most helpful in the present study. We have been able to co-ordinate their press releases of both Hodsoll and M-O items, then follow up contemporary reactions; likewise in Coventry, Manchester and Birmingham (chapter 12); cf. *Southampton Evening Echo*, February to May 1973. See also chapter 12, note 2 below.
10. Compare O'Brien, p. 643 for other aspects of armed service help in Southampton and elsewhere. The extent and limits of such support were never effectively communicated to the public, who often resented apparent preferential treatment for troops. See also Portsmouth below.
11. FR 529 of 19.12.40; the 'larder survey' cannot, alas, now be found, having been removed – with quite a lot more – from the files before the Archive reached the University of Sussex.
12. Opinion Research Centre poll in *Evening Standard*, 10.7.74 and *International Herald Tribune*, 11.7.74; Census results from *The Times*, 7.8.74.
13. The following paragraphs are mainly based on or taken from FR 603 of 9.3.41, largely the work of Mrs Priscilla Novy, then wife of another staff member, Henry Novy (see preface, note 1).
14. This notice (from FR 604:11) has been partially quoted in another context by Calder, 1951, op. cit., p. 321.
15. Data adapted from official sources and Knowles, op. cit., 1951: appendix A; cf. also A. Temple Patterson, *Southampton, a Biography*, 1970, p. 212 etc. The tonnage figures from local sources differ considerably from the German figures which are, of necessity, taken as a basis (preface, note 2; cf. Collier, op. cit., p. 506), largely because many bombs were dropped in the countryside, into water, otherwise off target, did not explode or went unrecorded into the ground. This difficulty applies to all blitz figures, only used within this study for overall comparison, not as absolutes.
16. Titmuss, op. cit., 1950, p. 300 etc. The 'poor law' aspects are more critically examined again in the striking case of Plymouth in our following text (7, 5 below).

17. Civic learning and non-learning are so important that they are summed up for the whole experience in chapter 10.

(2) *Portsmouth* (*the Nelson touch*)

1. FR 606 of 15.3.41; the other main Portsmouth reports are FR 559 and 850, the last supported by particularly rich raw material. Calder, op. cit., 1971, pp. 252–3 has used bits from these files. We also received valuable checking data in 1973–5 from Mrs Lee Chater of Havant, the City Librarian and the (wartime) Borough Engineer.
2. RM original, a very detailed survey, partly by regular interviewing on sample basis, then in its technical infancy and difficult under those conditions anyway. Devised by Mollie Tarrant, observer, Havant.
3. As well as our main observer with the mourning party another moved among the crowd. The 'mourner' had an embarrassing moment when a senior police officer started reading from a check-list by the graveside, rearranging mourners in sequence. But when there were six left he told them all to follow anyway.
4. *Portsmouth Evening News*, 14.5.41, 10.6.41, etc.
5. Compare also Rediffusion's blitz value compared with BBC at Plymouth.
6. None of these figures are intended to be taken as statistically exact, but they are from serious, detailed studies made for the Director of Naval Intelligence through 1941, only a part of the raw material for which remains in the Archive, alas. The main results were often communicated verbally direct, from the author to Admiral Godfrey; DNI was using them, at a high level, for purposes sometimes bordering on intrigue and inter-departmental warfare (as we now realize; see chapter 10), although also inspired by deep concern for sailor and docker families.

(3) *Three 'Defeats'?*

(no notes)

(4) *Bristol, in the south-west*

1. For bomb intensities see note 4 below. Main Bristol blitz sources are FR 529 of 19.12.40, FR 601 and 628; cf. FR 374, 397 and 465; also RM Box 290/36/F 3440, 3438. One pre-blitz whole-time observer in Bristol was Kathleen Box MBE who later transferred to the official post-war Social Survey, where she co-authored

important reports, including their two massive published studies of cinema audiences.

2. FR 295; cf. Calder, op. cit., p. 237. He also quotes Rev. S. P. S. Shipley's *Bristol Siren Nights*, Rankin (Bristol), 1943, p. 24; the priest's diary extract that night: 'God grant that it is going to be alright for us.'
3. O'Brien, op. cit., p. 643, cf. p. 175; see further in chapter 10.
4. It may be useful at this stage to summarize the overall bomb pattern for this and other provincial centres studied so far and to follow:

Place	*Number of major raids*	*Tonnage of HE*
London	71	18,800 (5149 after 14 Nov)
Merseyside	8	1957
Birmingham	8	1852
Clydeside	5	1329
Plymouth Devonport	8	1228
Bristol-Avonmouth	6	919
Coventry	2	818
Portsmouth	3	687
Southampton	4	647
Hull	3	593
Manchester	3	578

(No other place exceeded 450 tons)

(5) *Plymouth – Drake's Country* (*and Astor's*)

1. Lord Astor's Foreword to local reporter H. P. Twyford's *It came to our Door*, Underhill (Plymouth), 2nd edition, 1946.
2. FR 295 of 29.1.41; also FR 626 and 683; with an important follow-up in FR 903 of 6.10.41.
3. Harold Nicolson's diary for 7.5.41; see also Calder, op. cit., p. 248. Morrison's earlier refusal to accept bad reports is illustrated in chapter 10.
4. Churchill on BBC, 27.4.41; see also Calder, op. cit., p. 248. Lady Astor and Winston had a nasty little row that day (see items at note 6 below). A lively description (from memory) by Lord Ismay of Churchill similarly touring Bristol and Coventry is in *My Darling Clementine* by Jack Fishman, Allen and Unwin, 1974, p. 184.
5. John Hilton, *Rich Man, Poor Man*, Allen and Unwin, 1944, the text of the Sir Halley Stewart Lectures, 1938, with a foreword by Sir William (later Lord) Beveridge. An important little book, by a lucid

communicator who was then a great rarity, a full Professor (Cambridge) with working-class (Bolton) origins and an unacademic approach. Prof. Hilton, who died prematurely (1943), played a big part in setting up the MOI Home Intelligence under Mrs Mary Adams (see chapter 1). Titmuss, op. cit., p. 116 seems to double the number of poor families, though Hilton is not too clear on his statistics (110). See also Edna Nixon's *John Hilton*, Allen & Unwin, 1946, chapter XXVI; cf. in general now K. Coates and R. Silburn, *Poverty: The Forgotten Englishmen*, Penguin, 1970; V. Bogdanor and R. Skidelsky (eds), *The Age of Affluence*, Macmillan, 1970.

6. Several descriptions exist of Lord and Lady Astor under fire, two of them for the first, 'worst' night, 20 March, when the bombing started while she was at home recovering from the royal tea party earlier that afternoon. They sheltered at first, but emerged to cope with an incendiary; and presently for her ladyship to go to bed in her room with all the windows already blown out. There is a marvellous description in Christopher Sykes, *Nancy, The Life of Lady Astor*, Collins, 1972, of Nancy going into a public shelter where, to her, 'the morale seemed rather low'. She looked around, said nothing, then made a tour of the place turning cartwheels and somersaults. 'Are we down-hearted?', she cried, departing amidst roars of admiration. See also Maurice Collis, *Nancy Astor, an informal biography*, Faber, 1960; pp. 197–202. Hon. David Astor has kindly helped with other information on his mother, including direct contacts in Plymouth (1974). See also Rose Harrison, *Rose: my life in Service*, Cassell, 1975, especially chapter 8 ('A Family in Wartime'), a good account.

   More generally there is a diary by a distinguished (Prix Goncourt) Frenchman living in Plymouth through 1940–41, A Savignon, *With Plymouth Through the Fire*, Ouston (Hayle, Cornwall), 1972, which adopts almost the same heroic tone as local historian Twyford. Mr Peter Walke at the Open University is presently engaged in a thesis on the Astors in Plymouth, 1939–45. Robert Dougall, then a BBC commentator, was by chance in Plymouth for the second night of the 21 March attack and gives a good short account in his *In and Out of the Box*, Fontana, 1975, p. 106, including perceptive comment on 'ecstatic bliss' reactions and a fine Nancy Astor *faux pas* ('Plymouth Po').

8 THE INDUSTRIAL NORTH

(1) *Merseyside and the 'Martial Law'*

1. From FR 538, pp. 3–4 (slightly edited); other main sources FR 706, 711 and follow-up 1245 (May, 1942). Since the general pattern of blitzing civilians has been covered in previous chapters, this one emphasizes what happened differently further north – where there were differences (and these were *never* simply 'regional'). A special effort is also made to use other sources to support blitz descriptions as well as M-O's. All the rest of the top-ranking bombed towns are covered. Lesser sufferers such as Belfast, Cardiff, and Tyneside do not add to this record; data are available in the Archive.
2. FR 538; cf. Lancashire and Cheshire Community Council, AWM/MMS, 22.11.40.
3. M-O wartime by-election files, a good series; cf. Calder, op. cit., p. 335.

(2) *Manchester and Birmingham inland*

1. FR 538 of 6.1.41; cf. FR 609, 839; F 3368, 3464, 4170, etc.; with intensive 'Worktown' study of 1938–41 for Manchester area generally.
2. Research under Dr Tony Sutcliffe, Department of Economic History, Sheffield University, for his largely unpublished major study of civil defence administration in Birmingham (History of Birmingham Project, Research Papers Nos. 1, 2 and 3, which he generously gave us permission to use freely). See also H. J. Black's *History of the Corporation of Birmingham*, 1938–50 (Birmingham Corporation, 1957), and Dr Sutcliffe's published account (with Roger Smith), *Birmingham, 1939–70*, Oxford University Press, 1974. And further at appendix B.
3. Research and Experiments Department, Ministry of Home Security, in PRO, HO 199/453 (May 1942); and see further in chapter 10 following. Also *Birmingham Post*, 3.8.74 (Press Association release: CITY ON THE BRINK; cf. chapter 12).
4. M-O by-election files; and see chapter 11, 5–6.

(3) *The Clyde Towns*

1. FR 600 of 7.3.41; cf. Director of Home Intelligence file (Mary Adams gift in Archive), Box 296, correspondence 20.3.41; also Calder: 243,695 (using M-O). FR 607 on Clydebank and FR 631; follow-ups FR 778 and 932.

2. In Scotland the Regional Commission system was decentralized with five District Commissioners, one at Glasgow. This gave a closer tie to the major civic authorities, and worked better than, for instance, running south-east England from Tunbridge Wells (cf. O'Brien, op. cit., p. 675).
3. FR 600 goes into the background of Clydeside productivity in detail, and is taken up seven months later in FR 932 (cf. Titmuss, op. cit., pp. 313, 362; O'Brien, op. cit., p. 421).

(4) Not least, last – Hull

1. Herbert Morrison, *An Autobiography*, Odhams, 1960, p. 187. Archive FR 640 of 7.4.41, pp. 844 and 1213. Research and Experiments report at PRO, HO 199/453; Hull and East Riding Information Committee (no. 2), Special Report, April 1941. Titmuss, op. cit., p. 306, O'Brien, op. cit., pp. 412, 427; cf. Calder, op. cit., p. 250.

(5) Around Worktown (and the like): unblitzed

1. FR 404 and 856; RM Box 253.

(6) Provincial Epilogue: back to London . . .

1. FR 575, 587 and 588.
2. Frank R. Lewey, *Cockney Campaign*, Stanley Paul, 1944. See also literature in note 9 to chapter 5 above, and specifically Mosley, op. cit. p. 235 (on 29 December), Calder, op. cit., p. 240, Fitzgibbon, op. cit., p. 204. Also the rather difficult but thorough study of 10 May by Richard Collier, *The City that Wouldn't die*, Collins, 1959; also Robert Dougall, *In and Out of the Box*, Fontana, 1975, p. 105, 'London might have cracked'. Lewey's sometimes fulsome but critical book has a foreword by Clement Attlee, who describes the 'unconquerable spirit' of the Cockney and says that:

   'Wherever I went in those grim days I never found anything but cheerfulness.'
3. Raymond E. Lee, *The London Observer*, Hutchinson, 1972, p. 270; also Julian Huxley, *Memories*, Penguin, 1972, pp. 237, 243 on Churchill at the zoo and post-blitz 'sense of exhilaration' (cf. our 5 (4)).

9 MORALE QUESTIONS

1. Jaroslav Hasek, *The Good Soldier Svejk* (translated by Cecil Parrot), Penguin, 1975, p. 259.
2. Monthly M-O voluntary panel directives and M-O monthly